Edward Muntean
Food Analysis

Also of Interest

Microbiology of Food Quality.
Challenges in Food Production and Distribution During
and After the Pandemics
Hakalehto (Ed.), 2021
ISBN 978-3-11-072492-9, e-ISBN (PDF) 978-3-11-072496-7

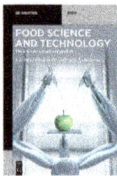

Food Science and Technology.
Trends and Future Prospects
Ijabadeniyi (Ed.), 2020
ISBN 978-3-11-066745-5, e-ISBN (PDF) 978-3-11-066746-2

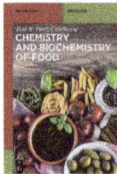

Chemistry and Biochemistry of Food
Perez-Castineira, 2020
ISBN 978-3-11-059547-5, e-ISBN (PDF) 978-3-11-059548-2

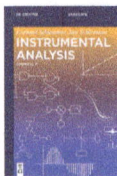

Instrumental Analysis
Schlemmer G, Schlemmer J, 2022
ISBN 978-3-11-068964-8, e-ISBN (PDF) 978-3-11-068966-2

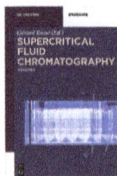

Supercritical Fluid Chromatography
Rossé (Ed.), 2018
Volume 1 ISBN 978-3-11-050075-2, e-ISBN (PDF) 978-3-11-050077-6
Volume 2 ISBN 978-3-11-061893-8, e-ISBN (PDF) 978-3-11-061898-3

Edward Muntean
Food Analysis

Using Ion Chromatography

DE GRUYTER

Author
Prof. Dr. Edward Muntean
University of Agricultural Sciences
and Veterinary Medicine of Cluj-Napoca
Calea Mănăştur Street 3-5
400372 Cluj-Napoca
Romania

ISBN 978-3-11-064438-8
e-ISBN (PDF) 978-3-11-064440-1
e-ISBN (EPUB) 978-3-11-064449-4

Library of Congress Control Number: 2022939182

Bibliographic information published by the Deutsche Nationalbibliothek
The Deutsche Nationalbibliothek lists this publication in the Deutsche Nationalbibliografie;
detailed bibliographic data are available on the Internet at http://dnb.dnb.de.

© 2022 Walter de Gruyter GmbH, Berlin/Boston
Cover image: D-Keine/E+/Getty Images
Typesetting: Integra Software Services Pvt. Ltd.
Printing and binding: CPI books GmbH, Leck

www.degruyter.com

Foreword

Food is defined as being "any substance, whether processed, semi-processed or raw, which is intended for human consumption and includes drink, chewing gum and any substance which has been used in the manufacture, preparation or treatment of 'food'" (Codex Alimentarius, 1985). Food products have to provide proper nutrients for consumers while being safe for them; this requirement can be fulfilled only with the support of food analysis – a set of techniques and methods delivering information about the chemical, physical, microbiological and sensorial attributes of food and raw materials involved in the food production. Since food analysis is concerned with the detection, identification and quantification of a wide area of substances of interest in food matrices, the food analyst has a major role in the whole food chain, being responsible for providing valid and accurate data on food composition. Moreover, the analyst must be able to communicate effectively with a diverse range of professionals (e.g. engineers, managers, inspectors, consumers, lawyers, police, etc.); hence, a good understanding of all analytical issues as well as technological and legal ones is beneficial. Data obtained by food analysis are critical to describe in an objective manner the properties of foods, to assure that they are safe, nutritious and desirable for consumers, being also necessary for a proper food quality management.

Numerous methods are currently used for food analysis: volumetric analysis, gravimetry, electrochemical methods, spectrometry, enzyme-based assays, etc. Among these, ion chromatography (IC) gained an important position in the last decades, becoming even the method of choice for many analytes in certain matrix types. As will be detailed later, IC is an analytical technique for the separation and determination of ionic solutes (inorganic cations, inorganic anions, low-molecular-weight organic acids, etc.). This book brings updated information about various applications of IC in food science, such as food quality control, food authentication and analysis of residues in certain food products. In addition, recent developments in instrumentation such as in-line eluent generation systems, capillary IC and combustion IC are also described.

By its content, the book is addressed to students interested in the subject but can be a real support also for researchers from related fields. Professionals in the food industry, as well as laboratory scientists, must have a detailed knowledge of analytical chemistry and instrumental analysis; they have to be able to integrate the scientific and technical information with a proper understanding of the context in which the data they provide are used. The textbook can also fit into the current concerns of both producers and consumers regarding healthy food products, knowing the close relationship between the state of the environment, the quality of agricultural crops and the quality of food products, since many ionic substances they contain are of major interest for human health. This book is intended to provide the necessary support for such an approach, offering a comprehensive review both for newcomers and more experienced in the field of IC. It is structured in six chapters,

https://doi.org/10.1515/9783110644401-202

in which effort has been made to maintain an appropriate balance between analytical principles, relevant technical details and applications.

Chapter 1 addresses the basic principles of chromatography, underlying some fundamental issues as well as several important concepts and a glossary of terms to facilitate a proper understanding of the next chapters. This chapter is intended for the new ones in the field; hence, the readers with proper knowledge of liquid chromatography can get over it.

Chapter 2 provides an insight into IC, highlighting the instrumental setup, with explanations for the principles behind the operation of each part of an IC system. It is intended to be a comprehensive introduction to IC, initiating the reader in the subject, giving also many relevant references (books, scientific papers, websites, etc.). An overview of the major systems' and columns' manufacturers on the IC market is included here to assist a laboratory manager in the first steps in the buying decision for implementing IC analysis in a certain infrastructure.

Chapter 3 reviews several applications in food science, targeting water, beer, wine, milk and dairy products. A wide range of directions are covered in this context, including food quality control, food safety issues and food authentication. Relevant data on IC analysis are tabulated and representative chromatograms can help the reader rapidly find a suitable application for a desired analytical context.

Chapter 4 covers several recent developments in the field (in-line eluent generation systems, capillary IC, combustion IC). Chapters 5 and 6 give some insights into practical approaches by detailing several sample preparation techniques (dilution, filtration, ultrafiltration, dialysis and solid-phase extraction) and addressing common troubleshooting issues.

Overall, the textbook is intended to serve as an organized resource for students, PhD students, researchers and food analysts. It highlights that IC can be even more powerful and efficient when more sophisticated equipment is available, while proper knowledge empowers the user to obtain relevant data from this. Fortunately, there is much information available to allow users to get the most out of this technique.

The author welcomes readers' comments, criticism and ideas for improvement of future editions of this book.

Special thanks to the editors – thank you so much for your patience and for your support!

Edward Muntean,
Professor,
University of Agricultural Sciences
and Veterinary Medicine Cluj Napoca, Romania

Contents

List of abbreviations

^{0}C	Celsius degree
atm	atmosphere
AU	absorption units
AUFS	absorption units full scale
α	Greek symbol for selectivity
CDS	chromatographic data system
CR-TC	continuously regenerated line trap column
DAD	diode array detector
DEAE	diethylaminoethyl
dil	dilute
ECD	electrochemical detector
EDTA	ethylenediaminetetraacetic acid
g	gram
FDA	Food and Drug Administration
GLP	good laboratory practice
GRAS	generally recognized as safe
HETP	height equivalent of a theoretical plate
HPAE	high performance anion exchange
HPIC	high-pressure (capillary) ion chromatography/high-performance (capillary) ion chromatography
HPLC	high performance liquid chromatography
i.d.	internal diameter
IC	ion chromatography
kg	kilogram
LC	liquid chromatography
LED	light emitting diode
LOD	limit of detection
LOQ	limit of quantitation
Λ	Greek symbol for wavelength
μ	Greek symbol for "micro"
M	molarity
mg	milligrams
mL	milliliters
m^3	cubic meters
M	molecular mass
MS	mass spectrometry
N	efficiency (number of theoretical plates)
N.A.	not applicable
p	pressure
Pa	Pascal
PAD	pulsed amperometric detection
PAR	4-(2-pyridylazo)-resorcinol
PDA	photo diode array
PEEK	polyetheretherketone
ppb	parts per billion
ppm	parts per million
ppt	parts per trillion

https://doi.org/10.1515/9783110644401-204

psi	pounds per square inch
PSDVB	polystyrene-divinylbenzene
PTFE	polytetrafluoroethylene
RI	refractive index
RID	refractive index detector
S	Siemens
S/ N	signal to noise ratio
SAX	strong anion exchanger
SEC	size exclusion chromatography
SCX	strong cation exchanger
SDVB	styrene divinyl benzene
SPE	solid phase extraction
t°	temperature in Celsius scale
t_R	retention time
UV	ultraviolet
V	volume
v/v, w/v	volume to volume or weight to volume
VIS	visible
WAX	weak anion exchanger
WCX	weak cation exchanger
WHO	World Health Organization

1 Introduction

Among liquid chromatographic methods, ion chromatography (IC) can be considered as one of the most valuable analytical tools, this being an affordable and advantageous mainstream analytical technique, able to provide a convenient determination of various analytes such as anions, cations, organic acids, carbohydrates, sugar alcohols, amines, aminoacids, aminoglycosides, proteins, peptides and glycoproteins. Comparing with techniques such as atomic absorption spectroscopy, spectrophotometry, titration or electrochemical applications using ion selective electrodes, IC provides the advantage of simultaneous determination of many ionic sample components. This technique is especially valuable for the analysis of ionic or ionizable substances that have little or no UV absorbance; its greatest utility is for simultaneous analysis of anions, for which there are no other rapid analytical methods.

Because of its high accuracy and reliability, IC is nowadays one of the most important methods for the determination of alkaline, alkaline earth and some transition metals as well as inorganic and organic anions from various matrices, offering an easy, fast, small sample volume demanding robust, highly sensitive and fit-for-purpose methodology for the routine determination in a large dynamic range, with relatively low running costs (Fritz and Gjerde, 2009; McGorrin, 2009; Michalski, 2016; Weiss, 2016).

IC is also advantageous by combining high sensitivity with a wide working range of analytes' concentrations, since nowadays, broad area of columns and detectors are commercially available and high selectivity can be obtained in many applications. The possibility to couple an ion-exchange separation with various detection options expands the IC applications to instances where analyte-specific detection strategies can provide high sensitivity and/or specificity. Hence, multispecies analysis can be accomplished with high precision and accuracy in a short time, in some cases IC providing a possibility to accomplish direct sample analysis (without sample preparation). Advances in sample preparation lead to many in-line approaches such as in-line ultrafiltration or in-line dialysis (Frenzel, 2002; Saubert et al., 2004, Steinbach and Wille, 2009; Wang, 2010; Xu et al., 2014), while in-line preconcentration lowered the detections limit to the ng/L range (Michalski, 2016; Xiong et al., 2014).

From the analysts' point of view, IC can be considered a versatile and flexible technique; it supports different system configurations and settings, column types, eluents, inline sample preparation techniques. Most systems are robust and simple to use, some of them having a high degree of automation and allowing large sample series (Fritz, 1987; Haddad and Jackson, 1990; Liu et al., 2015).

From the environmental point of view, IC is a safe and environmental-friendly technique; it typically uses dilute acids, alkalis or salt solutions as mobile phases (such as $NaHCO_3$, Na_2CO_3, KOH, NaOH, HNO_3 and methanesulfonic acid) and does not require organic solvents, hence it does not involve the purchase of costly substances and hazardous disposal of the waste eluent; the eluents can be disposed

https://doi.org/10.1515/9783110644401-001

after an eventual neutralization and/or dilution with water (D'Amore et al., 2021; Michalski and Pecyna-Utylska, 2020; Michalski and Pecyna-Utylska, 2021).

For a food analyst, IC is a valuable tool for analysis of various sample types, such as raw materials, end products, waste streams, production equipment cleaning solutions and process water; its wide dynamic range makes it applicable for the quantification of both trace contaminants and of major components (Haddad and Jackson, 1990; Paull and Nesterenko, 2013; Weiss, 2016).

1.1 Chromatography basics

This subchapter is intended to give a short introduction in chromatography for those less familiar with this subject, such that the concepts discussed later are easier to understand.

Chromatography is a large group of techniques used for separation and quantitation of target compounds from mixtures, based on their differentiated complex interactions with a **mobile phase** (a gas, a liquid or a supercritical fluid[1] that moves continuously into the system) and a **stationary phase** (an immobile solid or liquid deposited on an inert support). The mobile phase carries the components from the mixture along the stationary phase; as a consequence of the differential distribution of the components from the mixture between the two phases, a movement with different speeds of these occurs, resulting in their separation. If the separated components are passed through a detector, they generate an electric signal which is then recorded as a **chromatogram** – a graphical representation of the detector's response as a function of time (Snyder et al., 2011; Weiss, 2016).

The term "chromatography" originates from Greek (*chroma + graphein*), meaning "to write with colors," being proposed by the Russian botanist Mikhail Tswett to describe the colored zones obtained after the separations of plant pigments through columns (Tswett, 1901; Tswett, 1903; Tswett, 1906); chromatography has since developed into a valuable laboratory tool for the separation and identification of many classes of substances, maintaining its name despite most of them are colorless. The basic principle is still the same, the separations being carried out using two phases: a stationary one (the column packing) and a mobile phase (a fluid which moves through it).

The chromatographic methods can be classified according to different criteria (Snyder et al., 2011):

---

[1] A supercritical fluid is a gas brought to a temperature and pressure above critical values ($t\,^\circ > t_c\,^\circ$; $p > p_c$), a fluid with liquid-to-gas intermediate properties (with solubility and density similar to liquids, but gas-like viscosity and diffusivity).

- the separation mechanisms (e.g., adsorption, repartition, steric exclusion, affinity, ion exchange, ion pair formation, ion exclusion[2]);
- the mobile phase's aggregation state (Table 1, **liquid chromatography** – with liquid mobile phases, **gas chromatography** – using gases as mobile phase, **supercritical fluids chromatography** – using supercritical fluids);
- the used techniques: thin-layer chromatography, column chromatography;
- the final purpose of the separation (**analytical chromatography** – aims to identify and/or to quantify the components from the mixtures; **preparative chromatography** – aims to separate larger amounts of components from the mixtures for different further uses).

A common practice in the literature is to use acronyms such as "GC" for gas chromatography, "LC" for liquid chromatography, "IC" for ion chromatography, "HPLC" for high-performance liquid chromatography, "SFC" for supercritical fluid chromatography.

Table 1: Classification of chromatographic methods according to the nature of the phases.

Mobile phase	Stationary phase	Technique
Gas	Liquid adsorbed on a solid	Gas–liquid chromatography
	Solid	Gas–solid chromatography
Liquid	Liquid adsorbed on a solid	Liquid–liquid chromatography or partition chromatography
	Solid	Liquid–solid chromatography or adsorption chromatography
	Ion-exchange resin	Ion-exchange chromatography
	Macromolecular solid	Steric exclusion chromatography
Supercritical fluid	Solid	Chromatography with supercritical fluids

The movement of the components of the separating mixture carried out by the mobile phase along the stationary phase is called **elution**. The mobile phase is also called **eluent**, while the eluent leaving the chromatographic column is called **eluate**.

The chromatographic separation is mainly due to the differential relative affinity that the components of the mixture exhibit for the stationary phase. The components of the mixture move in the chromatographic system while they are in the mobile phase; those components that interact more strongly with the stationary

---

2 In some cases, chromatographic separations involve simultaneously more mechanisms (e.g., antigene–antibody, hydrophilic interaction liquid chromatography).

phase spend less time in the mobile phase and thus move more slowly through the system and vice-versa (Braithwaite and Smith, 2012; Dong, 2006).

In adsorption chromatography, the compounds from the sample mixture are reversibly bond to the stationary phase's surface by dipole–dipole interactions. Since the strength of these interactions is different for different compounds, the residence time in the stationary phase varies accordingly, thus achieving separation. Liquid–solid adsorption chromatography is most often used for the separation of polar, nonionic organic substances (Corradini et al, 1998; Snyder et al., 2011).

Partition chromatography involves a differential distribution of the compounds from a mixture between two immiscible liquids – one being the mobile phase, the other one a liquid coated or bonded to the surface of a solid support, representing the stationary phase (Robards and Ryan, 2021; Snyder et al., 2011). Depending on the polarity of the phases, one can distinguish in partition chromatography between:
- normal phase (NP) partition chromatography, in which the mobile phase is less polar than the stationary phase (used for the separation of polar organic compounds), and
- reverse phase (RP) partition chromatography, where the mobile phase is more polar than the stationary phase (used for the separation of nonpolar or weakly polar compounds).

Ion-exchange chromatography (IEC) is the best option for the separation of ionic compounds; the stationary phase consist of charged functional groups bonded to the surface of a macromolecular matrix, in contact with a mobile phase (an electrolyte), being able to participate in an ion-exchange process with charged species from the mobile phase. In fact, IEC is one of the earliest IC techniques, being used in the Manhattan Project for the separation of radionuclides (Settle, 2002). Ion pair chromatography is an alternative to IEC using stationary phases similar to that used in RP partition chromatography; it is based on the addition in the mobile phase of an ionic organic compound having the ability to form ion pairs with sample components of opposite charges, the resulting ion pairs behaving later as nonionic molecules that can be separated by RP partition chromatography (Haddad and Jackson, 1990; Weiss, 2016). Ion pair chromatography is advantageous because it enables the simultaneous separation ionic and nonionic analytes. Size exclusion chromatography (SEC) is used to separate analytes on the basis of their size and shape, the stationary phases being porous materials. In SEC, when a sample containing molecules having different sizes moves through the stationary phase, dissolved in the mobile phase, the molecules that are small enough enter the pores and are momentarily trapped and hence retarded, while larger molecules' interactions with pores are less likely, hence they are excluded and therefore carried out faster through the column, hence separation occurs because the difference in molecular sizes. Packing materials with different pores sizes are commercially available; they have to be selected depending on the sizes of molecules being separated (Berek, 2010; Mori and Barth, 1999).

Different modes allow for different interactions between the analytes and the stationary and mobile phases; from here results a wide area of possible adjustments of selectivity, since in liquid chromatography, the separation is based on the difference in the interactions of sample components between the mobile phase and the stationary phase. The mobile phase is a liquid that can dissolve the target compounds, while the stationary phase is hosted in most cases in a tubular device (a column), being specially designed for separating certain classes of substances; there are numerous types of commercial stationary phases, each of them being able to separate only a relatively small number of substances, for which they were designed. Since the stationary phase consists from small particles, tightly packed in the column, the mobile phase has to be pumped at high pressure in the system (Figure 1). As a compound moves through the column, it interacts with the two phases; hence, at any given time, a particular molecule is either moving in the mobile phase, or not moving at all in the stationary phase. The properties of each compound determine its behavior and retention and result in differential migrations of the components from the mixture. The different types of interactions that occur between the components of the mixture to be separated and the stationary phase (induction forces, dispersion forces, dipole–dipole interactions, hydrogen bonds, ionic interactions, etc.) are those that influence their migration velocity along the column and finally lead to the separation. In fact, the separation of the analytes from the sample mixture occurs because of the differences existing between their values of the distribution coefficients; as soon as the sample reaches the column, the analytes are distributed between the stationary and the mobile phases (Dong, 2006; Meyer, 2013).

The distribution coefficient of a component between the mobile and the stationary phases is the ratio between its concentration in the stationary phase (C_S) and in the mobile phase (C_M):

$$K_x = \frac{C_S}{C_M}$$

The values of K_x are determined by the physical and chemical properties of the respective compounds, by the natures of the stationary phase and the mobile phase and also by the working temperature. The separation of two components can be achieved only if they have different distribution coefficients; the difference between the distribution coefficients of the components of the sample to be analyzed is that which determines the difference between the migration velocities in the chromatographic system and finally the separation.

Since during a chromatographic process the eluent continuously moves on, clean amounts of eluent follows behind, then the distribution coefficient is restored by some of the sample components leaving the stationary phase into the eluent, while the portion of eluent that has moved on repartitions with the next particles from the stationary phase. This sequence occurs repetitively, as the analytes pass through the column. The sample components will therefore be retained in a differential

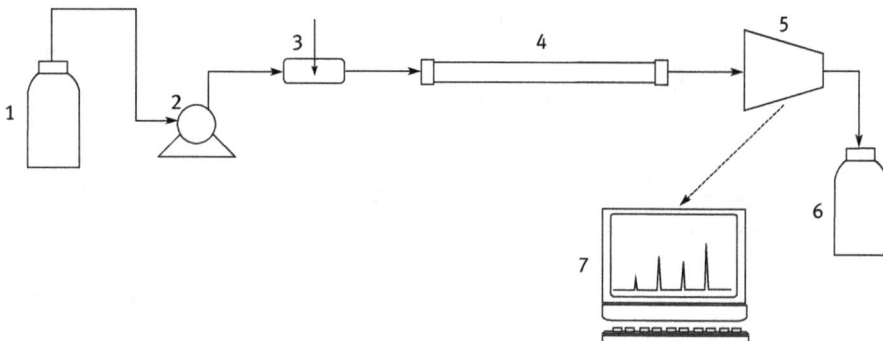

Figure 1: A simplified scheme for a liquid chromatographic system. 1, mobile-phase reservoir; 2, pumping system; 3, injector; 4, chromatographic column; 5, detector; 6, waste reservoir; 7, computer.

manner by the stationary phase and will have different migration rates, leaving the chromatographic column at different times, and then entering in an electronic device that monitors the eluent composition – a detector (Figure 1). The detector responds to a property of the analytes (e.g., conductivity, absorbance, fluorescence, refractive index) generating an electrical signal proportional with their concentration (Meyer, 2013; Robbards and Ryan, 2021).

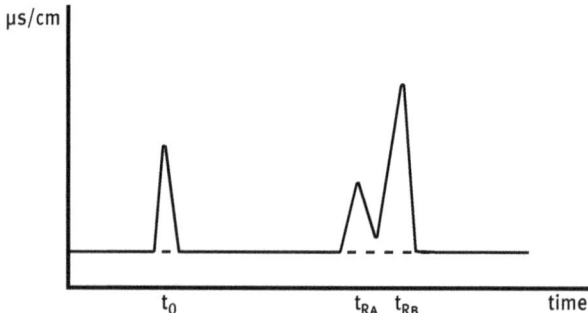

Figure 2: A model chromatogram corresponding to the separation of two components.

The graphical representation of the response of a detector over time is called **chromatogram**; under ideal conditions, each signal from the chromatogram (**peak**) corresponds to a component of the initial mixture. Figure 2 shows a model chromatogram for the separation of two compounds (in this case, conductivity as a function of time); the time corresponding to each peak (T_{RA} and T_{RB}) provides analytical information for identifying a component in the mixture in qualitative chromatographic analysis, while the peak height or the peak area can be correlated with the component concentration in quantitative chromatographic analysis (Dong, 2006; Snyder et al., 2011).

Chromatograms are sequences of peaks; ideally, these have Gaussian shapes, but in numerous cases they are not symmetrical (Figure 3).

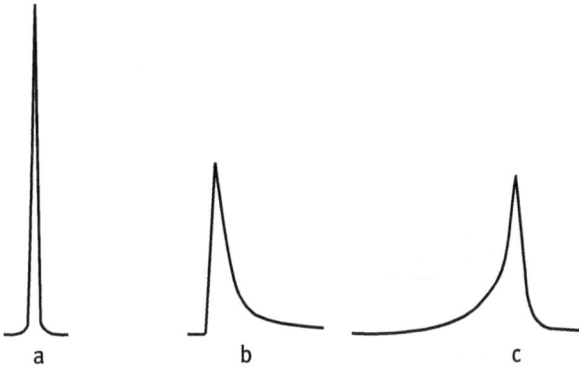

a   b   c

Figure 3: Peak profiles: a, symmetric; b, tailing; c, fronting.

The asymmetry factor is a measure describing the peak shape as a ratio calculated from the chromatographic peak by dropping a perpendicular from the peak apex on a horizontal line at 10% of the peak height (Blumberg and Klee, 2001); it is calculated by dividing the apex-to-back distance by the apex-to-front distance for the chromatographic peak, measurements being made at 10% of the maximum peak height (Figure 4), according to the formula:

$$S = \frac{CB}{CA}$$

where:
CA – the distance between the peak front and the perpendicular dropped from the peak maximum, measured at 10% of the peak height;
CB – the distance between the perpendiculars dropped from the peak maximum and peak end, measured at 10% of the peak height.

For a symmetrical Gaussian peak, the asymmetry factor is 1; as asymmetry increases, so does this factor (tailing peaks); values less than 1 denote fronting peaks (Figure 3).
    The tailing factor can be considered a particular case of the asymmetry factor, being a standard measurement in pharmaceutical applications; it is calculated in a similar manner as the asymmetry factor, but using measurements at 5% of the maximum peak height (Figure 5).
    The time a component requires for its elution (the time it spends in the mobile phase plus the time spent in the stationary phase) is called **retention time** (t_r); it is the time between injection and the appearance of the peak maximum (apex) of a certain component (Dong, 2006). The retention times are usually automatically determined by

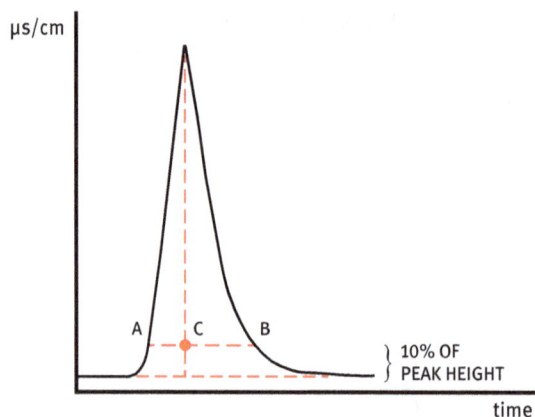

Figure 4: Establishing the value of the asymmetry factor.

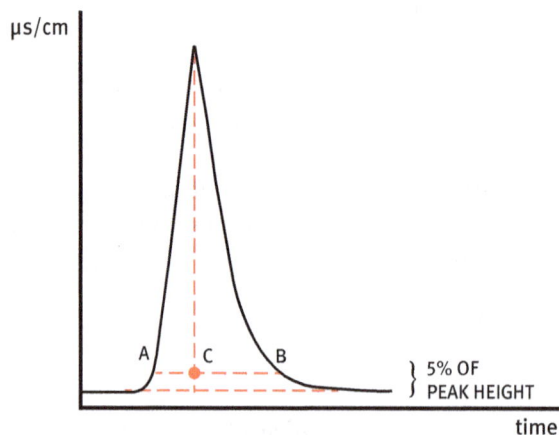

Figure 5: Establishing the value of tailing factor.

the chromatographic data system, being often included in the chromatograms and/or in the separation report (Figure 6).

The dead time (t_0) is the retention time of the components that were not retained in the system and which passed through the chromatographic system at the same rate as the mobile phase (Figure 7). It is important to monitor this parameter constantly, since this gives a feedback that the flow rate is the required one[3] (Meyer, 2013).

---

3 By multiplying the dead time with the flow rate, one can obtain the **void volume**; this is the volume (in mL) from the injection valve, through the tubing and the column, up to the detector. The void volume has a fixed value for a given system; it is affected by changing the column or the connecting tubing, but is not affected by factors such as temperature or a gradient, only by changes in flow rate.

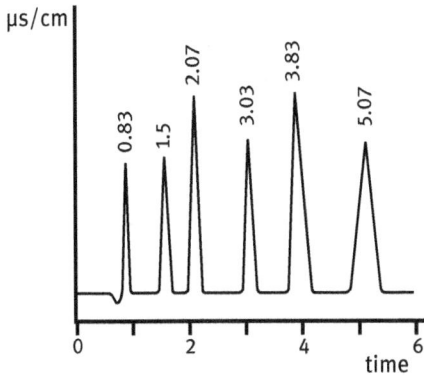

Figure 6: Chromatogram with peaks showing retention time.

The adjusted retention time (t_r') represents the difference between the retention time of a component and the dead time, being the time a component spends in the stationary phase (Blumberg and Klee, 2001; Snyder et al., 2011):

$$t_r' = t_r - t_0$$

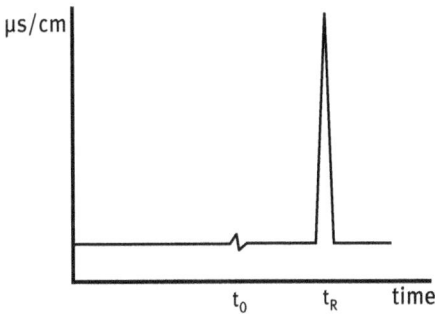

Figure 7: Model chromatogram corresponding to the separation of one component, highlighting the dead time and the retention time.

The degree of retention for a certain compound is given by the **retention factor (capacity factor, k'),** which is the ratio between the time that a sample component resides in the stationary phase relative to the time it resides in the mobile phase. It can be calculated using the formula:

$$k' = \frac{t_r'}{t_0} = \frac{t_r - t_0}{t_0}$$

where:
t_r – the retention time for the target peak
t_0 – the dead time.

In contrast with the retention time, the capacity factor for a given analyte is independent on the column dimensions and on the flow rate; low values for this dimensionless parameter are characteristic for poor separations, when the corresponding peak is located in the vicinity of the dead time, while large values lead to broader peaks and long analysis time. It has optimal values in the range 2–5 (Blumberg and Klee, 2001; Meyer, 2013).

Two or more substances can be separated only if they have different capacity factors, but to characterize the ability of a chromatographic system to separate them it is necessary to use a different metric. **Selectivity** characterizes the degree of separation between two consecutive components, representing a measure of the interaction between those components and the stationary phase. Selectivity is expressed by the **separation factor (α),** which is a measure of the relative retention for two substances; for two adjacent peaks from a chromatogram, it can be calculated as a ratio between the adjusted retention times of those peaks or the corresponding retention factors (Figure 8), according to the following relationship:

$$\alpha = \frac{k'_B}{k'_A} = \frac{t_{rB} - t_0}{t_{rA} - t_0}$$

Since B is more strongly retained ($k'_B > k'_A$), the selectivity is always higher than one and the higher the separation factor, the better the separation; however, when the separation factor increases, the time required for the separation increases too, hence separation factors around 1.5 are aimed (Dong, 2006; Meyer, 2013). If two analytes have the same chromatographic behavior on a certain system, they cannot be separated and coelution occurs ($\alpha = 1$)

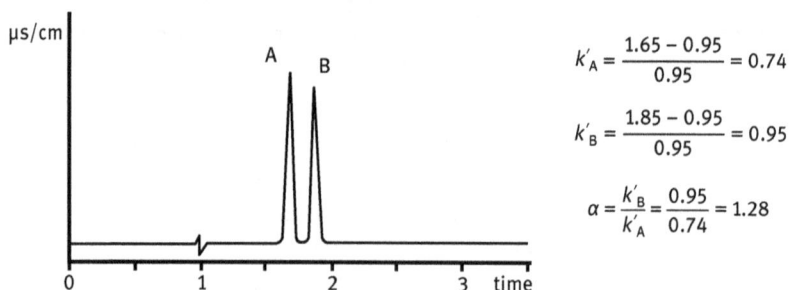

$$k'_A = \frac{1.65 - 0.95}{0.95} = 0.74$$

$$k'_B = \frac{1.85 - 0.95}{0.95} = 0.95$$

$$\alpha = \frac{k'_B}{k'_A} = \frac{0.95}{0.74} = 1.28$$

Figure 8: Model chromatogram corresponding to the separation of two components, highlighting the calculation of selectivity factor.

Because k' is constant for a given chromatographic system, so is selectivity. In fact, selectivity is a function of the interaction between the mobile phase and the stationary phase in equilibrium with the analytes, hence to change selectivity it is necessary to change the stationary or mobile phases; changing the mobile phase is the most affordable option, with the greatest changes that can be achieved, since different mobile phase interact with the stationary phase and with the sample in different ways (Robbards and Ryan, 2021; Snyder et al., 2011).

Resolution is a measure for characterizing the ability of a system to separate two successive peaks, as well as the quality of a separation; it can be calculated taking into account both the difference between the retention times and their geometry, being the ratio of the distance between peak maxima and the average baseline width of the considered adjacent peaks:

$$R = \frac{2 \cdot (t_{rb} - t_{ra})}{w_a + w_b} = \frac{1,177 \cdot (t_{rb} - t_{ra})}{w_{0.5a} + w_{0.5b}}$$

where:

t_{rB} and t_{rA} – retention times for the compounds B and A (Figure 9);

w_a and w_b – the corresponding widths at the bases of the peak (given by the intersection points of the inflectional tangents with the baseline);

$w_{0.5}$ – the peaks' width at half-height.

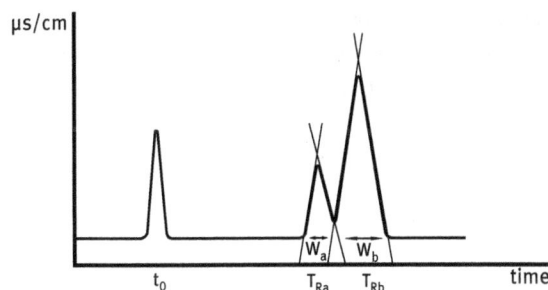

Figure 9: Model chromatogram corresponding to the separation of two components, highlighting the peaks' widths.

Resolution characterizes the quality of a separation since it takes into account both the relative position of the targeted peaks and their baseline width. A good resolution is obtained when the retention times corresponding to two consecutive peaks is large in relation with their baseline width.

Resolution depends on the difference between the distribution coefficients of the components. It is expressed as a number and values of 1.5 or greater are considered for baseline resolution for two adjacent peaks of equal size; a value of 1 is the minimum for a qualitative separation and 1.2 to allow quantitation, while large values of the resolution can lead to long analysis time. In practice, it is necessary to calculate the value of resolution especially when optimizing separations, when the objective is maximizing resolution while the duration of overall separation is as low as possible.

Resolution can be maximized by using longer columns, columns with smaller particle diameters for stationary phases and optimum flow rates, but there are drawbacks to consider:

- with longer columns and higher flow rates the mobile phases' consumption increases,
- by reducing the particle diameter one can experience an increase of the pressure in the system, which may exceed the maximum allowable pressure.

The most efficient way to improve resolution is by increasing selectivity (by changing columns to a different stationary phase and/or by changing the mobile phase's composition); in fact, in any chromatographic separation one have to select proper stationary and mobile phases, so that an optimal value for resolution is ensured (Dong, 2006; Meyer, 2013; Snyder et al., 2011).

For expressing the efficiency of a chromatographic separation, the separation process was related with the theory of distillation (Martin and Synge, 1941) and the concept of theoretical plate was introduced. A **theoretical plate** is considered a section of a chromatographic column in which reversible distribution equilibrium of a component between the mobile phase and the stationary phase occurs (Snyder et al., 2011).

The number of theoretical plates (N) is a measure of the efficiency of a chromatographic system, easy to calculate according to the formula:

$$N = 5.54 \cdot \left(\frac{t_r}{w_{0.5}}\right)^2 = 16 \cdot \left(\frac{t_r}{w}\right)^2$$

where:
t_r – the retention time of a chromatographic peak;
w – the peak width at the base (baseline width);
$w_{0.5}$ – the peak's width at half-height (Figure 10).

Since the retention times and peak widths are measured in time units, N is expressed as a dimensionless number; it characterizes the quality of a column packing and mass transfer phenomena that occurs during chromatographic separations – the larger it is, the more complex sample mixtures can be separated with the corresponding column.

Efficiency is affected by several parameters, from which:

- particle size – the smaller the particles from the column packing, the higher the efficiency;

- temperature – the higher the temperature at which separation occurs, the faster mass transfer become, hence peaks are sharper and the corresponding retention times smaller;
- mobile phase viscosity – the lower the viscosity, the more efficient the mass transfer process is; this factor is closely related with temperature, since viscosity decreases when temperature increases;
- flow rate – there is an optimum flow rate at which efficiency is highest (at very low and very high flow rates, the efficiency is poor).

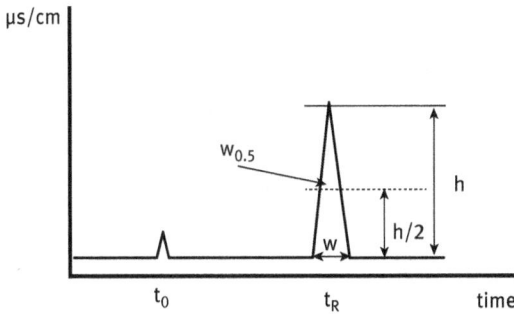

Figure 10: Model chromatogram corresponding to the separation of one component, highlighting the peak's widths for establishing the number of theoretical plates.

Knowing the number of theoretical plates for a column, the resolution can be established using the "resolution equation":

$$R = \frac{\sqrt{N}}{4} \cdot \frac{\alpha - 1}{\alpha} \cdot \frac{K'_2}{1 + k'_2}$$

where:
N – the number of theoretical plates;
α – the separation factor,
k'_2 – the capacity factor of the later eluted peak.

According to the previous formula, one can improve the resolution for a certain separation by increasing:

- the k' value (by changing the composition of the mobile phase);
- N (by using longer columns and/or columns packed with smaller particles – but the overall analysis duration increases as well);
- the selectivity (by changing the composition of the mobile phase and/or by using a different column).

It is important to check the plate number of a column when it is first received and installed on the system. The manufacturers provide a *test chromatogram* for each column showing the separation of a mixture of compounds with a specified set of conditions (flow rate, mobile phase, temperature, etc.). One should be able to get a close separation to that reported in the test chromatogram, and the plate numbers for different peaks should be also close to the reported values. The test chromatogram should be saved and used later, for troubleshooting purpose.

The height equivalent of a theoretical plate is considered to be the length of a column corresponding to a theoretical plate, the range in which the distribution equilibrium of a component is achieved between the mobile phase and the stationary phase. For a column of length L, it can be calculated using the formula:

$$H = \frac{L}{N}$$

However, since the chromatographic separations involve continuous sequences of equilibrium, it is not possible to individualize a separation, hence H and N have only an abstract significance (the column packings are homogenous; there are no real plates in a chromatographic column). Despite this, the number of theoretical plates is used frequently for the comparison of columns' efficiency.

Knowing the number of theoretical plates and the resolution, one can calculate **the peak *capacity*** (the number of equally well-resolved peaks that can be fit in a chromatogram-n); for a resolution of 1, it is given by the equation (Majors and Hinshaw, 2013):

$$n = 1 + 0.25.\sqrt{N}.\ln(1 + k_n)$$

where:
N – the number of theoretical plates,
k_n – the retention factor for peak n.

1.1.1 Qualitative analysis

After performing a certain IC separation, the support needed for obtaining analytical results is the chromatogram. In an ideal case, the number of the peaks from a chromatogram corresponds to the number of analytes from the sample (except the first one, which corresponds to the dead time); when complex samples are analyzed, it is possible to have **coelutions** (cases in which two or more substances have similar chromatographic properties and elute at a same retention time), hence the number of the analytes is at least equal with that of chromatographic peaks.

For qualitative purposes, the target parameter of the separation is the retention time of the analytes: analyte's identification in IC is usually accomplished by matching the retention time to the retention time of a pure standard run under identical conditions, hence it is compulsory to have reliable reference compounds for the substances of interest. The simplest qualitative approach is to compare the chromatogram of standards with the chromatograms of samples, both being obtained in identical chromatographic conditions (Haddad et al., 1991; Weiss, 2016).

In principle, if a separation is repeated under identical conditions, each analyte will elute from a chromatographic column at a certain retention time, which can be used later for identification purposes. Because the retention time depends on the

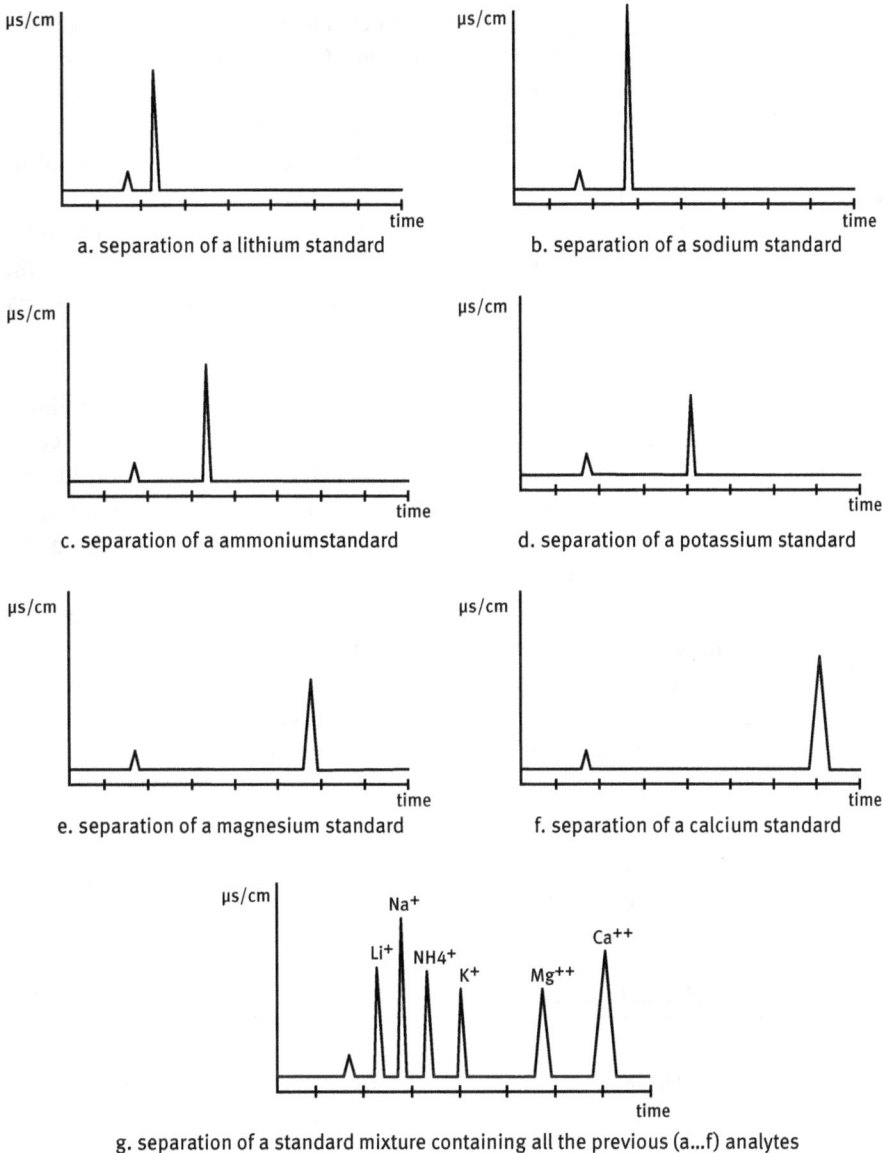

a. separation of a lithium standard

b. separation of a sodium standard

c. separation of a ammoniumstandard

d. separation of a potassium standard

e. separation of a magnesium standard

f. separation of a calcium standard

g. separation of a standard mixture containing all the previous (a...f) analytes

Figure 11: Chromatograms of reference solutions (a–f), each showing a single peak and of a mixture of standards, with peak IDs' revealed (g).

working conditions, it is imperative to maintain them both for the separation of the reference substance and for the samples.

As mentioned before, for performing qualitative analysis, reference analytes are necessary; by injecting them individually, using the same conditions that we use to separate them in different samples, one can obtain the retention time for each. For instance, in the case of cation analysis, by injecting reference solutions containing each of the cations Li^+, Na^+, K^+, NH_4^+, Mg^{2+} and Ca^{2+} alone, the corresponding chromatograms (Figure 11a–f) reveal the retention times (2.34 min for Li^+, 2.89 min for Na^+, 3.38 min for NH_4^+, 4.15 min for NH_4^+, 5.75 min for Mg^{2+} and 7.22 min for Ca^{2+}); by injecting a mixture containing all the abovementioned cations, the resulted chromatogram will have the identities of analytes automatically assigned by the chromatographic data system (Figure 11g).

When carrying out IC separations in order to establish the identity of individual peaks in a sample, one has to keep all experimental conditions constant during runs (same column, same mobile phase, same flow rate, same temperature, etc.) as they were used for standards. Comparison of the retention times with those obtained for standards reveals the peaks' identities.

In many cases, one may find extra peaks in the chromatogram besides those corresponding to the targeted analytes; they indicate that there are additional compounds in the sample, needing eventually a supplementary analytical effort to process them. In other cases, one or more peaks corresponding to the targeted analytes are missing; one can then report that the concentration(s) of that analyte(s) in the sample is below the detection limit. If comparing the retention times from chromatograms in Figure 12a and b, one can conclude that the sample contains analytes 1, 2, 4 and 5 (they have the same retention times) and not analyte 3 (with no corresponding peak), but two more unassigned compounds are also present. Such comparisons are actually accomplished by chromatographic data systems, which assign peak identities for retention times based on a certain analytical protocol.

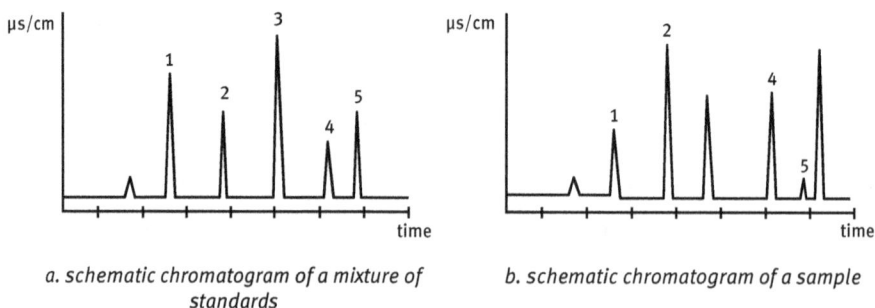

a. schematic chromatogram of a mixture of standards

b. schematic chromatogram of a sample

Figure 12: Qualitative analysis based on the comparison of retention times of pure standards (a) with those from a sample (b).

The retention times are usually not affected by peak sizes or the amount of an analyte in the sample; however, in certain cases, when the amount of samples are too large, overloading occurs and changes in both peak shapes and retention times occur. Peaks' shapes can also change as a result of interfering compounds with close retention times. Such cases, again, eventually need a supplementary analytical effort to process them.

For quality control purposes, it is advisable to perform a separation of standard compounds on a daily basis and to monitor the retention times for each peak; small variations (~few hundredths of a minute) are normal, while larger ones can be a symptom of some problems affecting the analytical procedure.

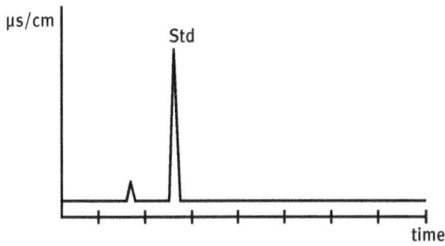
a. Schematic chromatogram of a pure standard

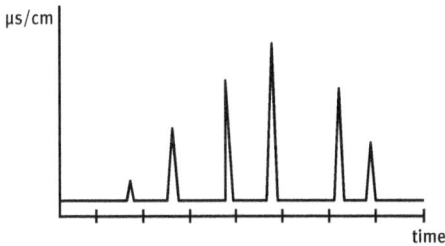
b. Schematic chromatogram of a sample

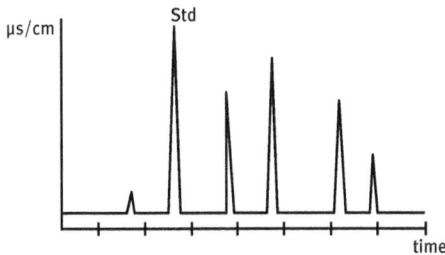
c. Schematic chromatogram of a sample + standard

Figure 13: Qualitative analysis based on co-chromatography: positive identification of a peak (the sample contains the analyte "std").

A qualitative approach which requires more sample and working time is that of **co-chromatography**[4]. To perform it, in a given sample which is first subjected to IC analysis as it is, then a known standard is added and the sample with the added standard is then injected again in the IC system. If the injected standard is present in the sample, then the corresponding chromatographic peak will exhibit a bigger area (Figure 13), otherwise the added standard will cause a new peak in the chromatogram (Figure 14).

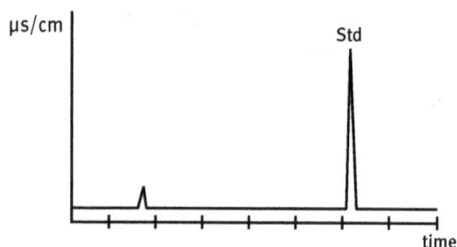

a. Schematic chromatogram of a pure standard

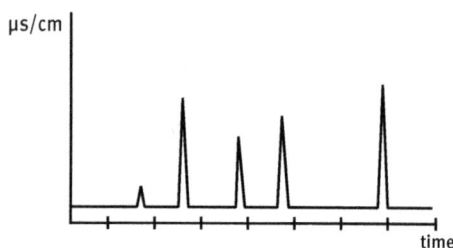

b. Schematic chromatogram of a sample

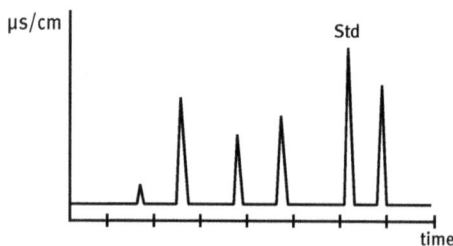

c. Schematic chromatogram of a sample + standard

Figure 14: Qualitative analysis based on co-chromatography: negative identification of a peak (the sample does not contain the analyte "std").

---

4 *Co-chromatography* is a technique in which a sample is mixed together with one or more known reference compounds, in the expectation that the relative behavior of the unknown and reference substances will assist in the identification of the unknown(s).

Supplementary identification criteria can be obtained using detection techniques such as mass spectrometry (which provides the m/z ratio) or photodiode array detection (which provides the UV-VIS spectrum). Hence, certain analytes, which shows absorbance in UV-VIS light can be identified based on the spectral characteristics when using a photodiode array detector; this detector allows spectral analysis of the acquired chromatographic data and by comparing the acquired spectra with those from the spectral library of the instrument, one can have an additional identification tool for qualitative analysis. By interfacing an IC system with a mass spectrometer (such as in IC-MS systems), the user can obtain the maximum qualitative information, since this approach gives the rations m/z for each peak.

1.1.2 Quantitative analysis

Quantitative analysis deals with the measurement of concentration for different analytes in the sample, concentration meaning the amount of the analyte in a certain quantity or volume of sample. Concentration is commonly expressed either as weight per volume or as weight per weight, using as units gram and liter (together with their corresponding multiples and submultiples).

Quantitative chromatographic analysis is based on the relationship between the concentration of an analyte and the peak area or peak height it generates in chromatogram. By using the chromatograms recorded after injecting reference standards that contain known amounts of the analytes, using a proper instrument and reproducible analytical conditions, the size of the chromatographic peaks are related with the injected amount of each, hence it is possible to determine how much of each analyte is contained in the analyzed samples (Dong, 2006; Weiss, 2016).

Two different size measurements are used in practice:

- **peak area** – since the area under the peak is directly proportional with the mass of analyte injected (the area of a peak being computed by the chromatographic data system through specialized algorithms);
- **peak height** – used when peaks' width and asymmetry are constant from run to run, since the peaks' height are directly proportional to area (peaks' height are also computed by the chromatographic data system), this approach being a more convenient one in certain cases, such as those with noisy baselines.

In order to provide quantitative results, an IC system has to be calibrated for the analytes of interest. **Calibration** is a procedure consisting in running a series of chromatographic separations with different known analyte concentrations that bracket the expected concentrations of analytes in the samples; finally, peak areas or peak heights are then related with the used concentrations (Figure 15).

The external standard calibration method is usually applied for quantitative analysis; for this, it is necessary to prepare several solutions with known concentration

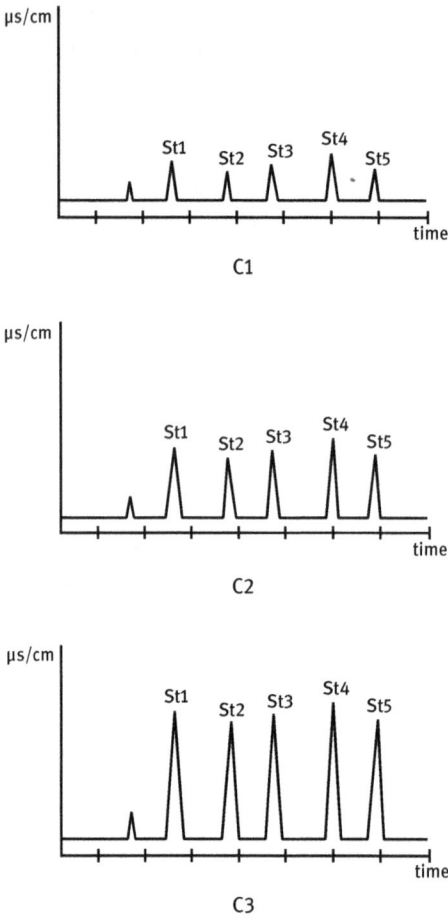

Figure 15: Schematic chromatograms corresponding to increased concentrations of analytes (C3 > C2 > C1).

of reference compounds,[5] then to perform the chromatographic separation for each under identical conditions as for the sample (Dong, 2006; Meyer, 2013). By measuring the areas or the heights of the resulted peaks and relating them with the concentrations, a calibration function is established for each analyte which can be described by an equation obtained from the measured points, obtained with the aid of linear regression analysis:

---

[5] The easiest approach is to buy commercially available standard s (e.g., Dionex seven anion standard, Dionex six cation standard – from Thermo Fisher Scientific) and to dilute them accordingly.

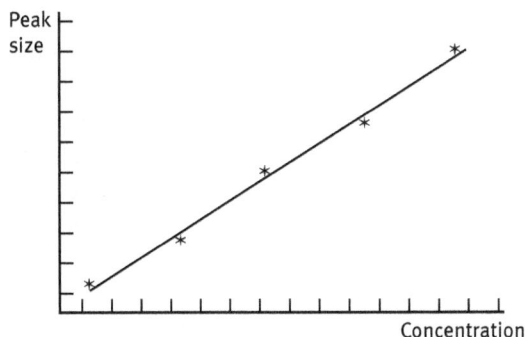

Figure 16: Calibration curve – a plot of peak size (area or height) as a function of the concentration of analyte injected.

$$C_i = m.A_i + n \quad \text{or} \quad C_i = m.H_i + n$$

where:

m – the slope of the calibration line,
n – the intercept of the calibration line,
C_i – the concentration,
A_i – the peak area,
H_i – the peak height.

The chromatographic data system performs an appropriate least-squares fit to the calibration points, obtaining the slope and the intercept of the calibration line; when a sample is run, the area of the analyte peak is measured and the corresponding concentration is calculated based on the established slope and the intercept. The linearity of the calibrations is assessed by calculating the square of the correlation coefficient, R^2: the closer is this to one, the better the fit.

The graphical output obtained in this way is the best-fit line established for the calibration points, being called **calibration curve** (**calibration line**). Calibration curves are obtained by plotting peak sizes on the y-axis versus sample concentration on the x-axis for the series of samples with known concentrations; linear calibration plots allow the simplest approach to quantitation and yield the best analytical results (Figure 16).

For obtaining accurate results, several conditions have to be met, such as:

- the standards used for calibration must have a high and well-known purity;
- the chromatographic conditions for both calibration and consecutive determinations must be identical (column, mobile phase, flow, temperature, injection volume, etc.);
- the peaks' magnitude from the sample under investigation should lie between those corresponding to the lowest and to the highest calibration standard, since the calibration function is defined only within the investigated concentration range.

The slope of the calibration curve is also referred as **sensitivity** (the ratio between the change in the signal and the change in the analyte concentration); the larger the increase of the signal with a change in the concentration, the bigger the sensitivity.

If the peak area (or peak height) for the analyte of interest is outside of the calibration range for a certain sample, that sample has to be diluted or concentrated, then a new chromatographic separation is necessary; in the final calculation on has to consider the sample preparation achieved. The extrapolation of the calibration line has to be avoided.

The "single point calibration" (or "one point calibration") is still used in many cases, despite is not recommended at all; it is based on injecting a standard solution of the substance being analyzed of about the same concentration as that in the sample (Jackson et al., 1995). The concentration of the analyte in the sample is calculated as follows:

$$C = \frac{P_{as}}{P_{ast}} \cdot C_{st}$$

where:
C – the concentration in the analyzed sample,
C_{st} – the concentration of standard solution,
P_{as} – the peak area for sample,
P_{ast} – the peak area for standard.

This approach can lead to inaccurate results either because an error in establishing any of the involved terms can lead to wrong results, or if the relationship between peak area and concentration is not a linear one. However, the "one point calibration" can be effective when the range of the determined concentrations is small and close to that of the used standard.

Internal standard calibration can be a valuable alternative to consider for external calibration, but this requires a supplementary reference substance (Hewavitharana, 2009). This method involves the addition to the sample to be analyzed, prior to injection, of a known amount of a substance ("**internal standard**") which has to fulfill simultaneously the following conditions:
-   it must be a high-purity substance,
-   it should have chromatographic and detection characteristics similar to those of target analytes from the sample,
-   it should be eluted as a separate peak in the chromatogram;
-   it should not interact chemically with other components of the sample.

The internal standard has to be added in such a proportion to the sample to achieve a concentration similar to that of the component(s) of interest. The method is remarkable precise, eliminating experimental errors due to injection.

1.2 Glossary

This section is intended to be a fast reference to those who have limited or no experience in chromatography, providing a basis from which the analyst can apply a concept (Corradini et al., 1998; Majors and Hinshaw, 2013; Majors and Carr, 2001; Meyer, 2013; Snyder et al., 2011; Weiss, 2016).

Absorbance: a measure of the amount of light absorbed by the eluent passing through the detector cell, established by means of a UV-VIS detector or a photodiode array detector and expressed in absorption units (AU)

Absorption spectrum: a graphical representation of the amount of electromagnetic radiation absorbed by a sample, usually presented as intensity versus wavelength

Accuracy: the closeness of agreement of the results of the analysis with the accepted value of a sample having known concentration of the analyte

Amphoteric ion-exchange resin: ion-exchange resins that have both positive and negative ionic groups

Analyte: a compound of interest to be analyzed

Anion-exchange chromatography: ion-exchange chromatography that uses stationary phases with functional groups designed to separate anions

Asymmetry factor: a measure of how much the shape of a peak deviates from the symmetrical bell shape

Atmosphere (atm): unit of pressure, 1 atm = 101.325 Pa = 1.01325 bar = 14.696 psi.

Autosampler: a module that automatically injects samples from sealed vials in the chromatographic system

Bar: unit of pressure, 1 bar = 100.000 Pa = 0.98692 atm = 14.503 psi

Base peak width: the distance between the points of intersection of the slope tangents with the baseline in a chromatogram

Baseline: the response of a detector on the data system display when only mobile phase is passing through it

Calibration: procedure that allows quantitative analysis, based on using standard compounds which are separated in a certain chromatographic context, same as used for sample analysis

Calibration plot: a graphical representation of peak sizes (height or area) on the y-axis versus sample concentrations on the x-axis, using a series of analytes with known concentration; also called calibration line if the plot is linear.

Capillary column: column with inner diameters less than 0.5 mm

Capillary tubing: tubing which connect various parts of a chromatograph

Cation-exchange chromatography: ion-exchange chromatography that uses stationary phases with functional groups designed to separate cations

Check valve: a device from the pumping system that allows flow of the liquid stream in only one direction

Chromatogram: the graphical output of the detector signal versus time obtained from the chromatographic process

Column packing: the particulate material packed inside a chromatographic column (also called stationary phase)

Column switching: technique using two or more columns connected by switching valves, for better chromatographic separations or for sample cleanup.

Counterion: an ion in solution which displaces the ion of interest from the ionic site in an ion-exchange process.

Dead time: (t_0) the time it takes for mobile phase or unretained sample compounds to pass through the chromatographic column and reach the detector

Dead volume: of the system: the volume of the tubing that connects the injector, the column and the detector

Degassing: the removal of dissolved gasses from the mobile phase, usually using vacuum

Detector: an electronic device which monitors continuously a certain property of the eluent leaving the chromatographic column, generating an electric signal which is proportional with the concentration of the analytes

Drift: a regular change in the baseline that is not perfectly horizontal, but rises or decreases during the separation time

Efficiency: the ability of chromatographic columns to produce well resolved and narrow chromatographic peaks

Eluate: the combination of mobile phase and solute leaving the chromatographic column

Eluent: the mobile phase used to perform a separation; the liquid that transports the sample through the IC system and contributes to the selectivity of the separation. It can be a solution of one or more substances in water; its ionic strength, pH, temperature, flow rate influence the selectivity of the separation.

Elution: the process of passing a mobile phase through a chromatographic column

End-fitting: the fitting that closes off the end of a chromatographic column and connects to the tubing through which the mobile phase enters and leaves the column

Fittings: the connectors that join tubing, columns and various modules together in a chromatographic system

Flow cell: the part of a detector through which the eluate passes, where the analytes are detected

Flow rate (_F_): the volumetric rate of flow of a mobile phase through a column, expressed usually in mL/min

Frit: a porous metal filter that is used to contain the column packing inside the column, or to protect other parts of the chromatographic system; its porosity can vary from 0.5 μm (for in-line filters for column protection) to 10 μm (for inlet filters attached to the pump inlet line)

Gradient: a programmed change in the mobile phase composition in time for eluting more highly retained analytes, intended to decrease the separation time

Guard column: a small column placed before the analytical column, which protects the analytical column from contamination by sample particulates and strongly retained species; it is usually packed with the same material as that from the analytical column and has the same inner diameter, but is much shorter

HPLC: high-performance liquid chromatography or high-pressure liquid chromatography, an application of liquid chromatography which uses instrumentation to allow complex separations and analyses to be carried out in short times

Hyphenated techniques: combinations of two or more techniques to separate and detect analytes (e.g., IC–mass spectrometry (MS), IC–MS–MS)

Inlet: the "entrance" port of a chromatographic column, the place in which the mobile phase and the sample enter

Inlet filters (sinkers): 10µm frits attached to the inlet lines in the mobile phase reservoirs

In-line filter: a frit held in a frit-holder mounted after the injector that prevents particulate matter from damaging the chromatographic column

Injector: a device used to introduce a precise volume of sample into the chromatographic system

Interference: an unwanted peak that overlaps one of the desired peaks in a chromatogram

Ion-exchange capacity: the number of ionic sites on a chromatographic column packing that can participate in the exchange process, expressed in milliequivalents per gram

Ion-exchange chromatography: a mode of chromatography in which ionic substances are separated on cationic or anionic sites of the packing, in which the sample ions, usually with a counterion, will exchange with ions already on the ionogenic group of the packing, retention being based on the affinity of different ions for the sites, as well as on other parameters such as pH, ionic strength and counterion type

Isocratic: elution using a constant composition of the mobile phase during the chromatographic run

Limit of detection (LOD): the smallest concentration of an analyte from a matrix which gives rise to the conclusion that the analyte is present and can be reliably distinguished from the analysis of a matrix that does not contain the analyte. LOD is one measure of sensitivity, usually defined in analytical laboratories as "3 x noise" (Armbruster and Pry, 2008); it can be calculated using the formula:

$$LOD = 3 \cdot \frac{s}{m}$$

where:

s – the standard deviation of the response estimated by the standard error of y intercept of the calibration curve,

m – the slope of the calibration curve.

Limit of quantitation (LOQ): the smallest concentration of an analyte that can be reliably quantified (with acceptable accuracy and precision) using a given system. LOQ is another measure of sensitivity, usually defined by the analyte concentration that gives a signal-to-noise ratio of 10 (Armbruster and Pry, 2008); it can be calculated using the formula:

$$LOQ = 10 \cdot \frac{s}{m}$$

where:

s and m – have the same significance as for LOD.

Linear range: concentration range over which a calibration is linear

Linearity: refers to the agreement of the results of the method used to determine the quantity of the analyte in samples of known concentrations with those concentration values

Loop: a piece of tubing in a sample injector that holds the sample to be injected and determines its volume

Matrix: refers to the sample in which the target analyte(s) is(are) to be analyzed (everything from a sample, other than the analytes).

Matrix effect: changes in the response as a result of certain component(s) which are present in the matrix

Method development: a process for fine tuning a separation to obtain a reproducible and robust determination, involving sample preparation, selection of an appropriate combination of stationary phase/eluent composition/column temperature/detection and data processing parameters that provides an adequate determination

Method validation: a process of testing a method to show that it performs to the desired limits of precision and accuracy in retention, resolution and quantitation of the sample components of interest

Mixed-bed column: combination of two or more stationary phases in the same column

Mixed-mode separation: a separation that occurs in a mixed-bed column caused by the retention and selectivity provided by a dual-retention mechanism

Mobile phase: a liquid phase (in IC) that carries the sample components through the chromatographic column, which may interact with both the analytes and the stationary phase

Mobile-phase additive: a substance which is added to the mobile phase in order to change the interactions between the analytes and the stationary phase or to prevent the bacterial contamination (e.g., acetonitrile, acetone, isopropanol and methanol)

Multidimensional chromatography: a combination of two or more chromatographic columns or chromatographic techniques designed to provide a better separation; it can be accomplished usually by using a switching valve which direct a particular fraction eluted from one column into a second one or system that has a different separation characteristic. It is a useful approach for sample cleanup, increased resolution, increased throughput and increased peak capacity.

Noise: a random change of the analytical signal around a certain value due to some causes which cannot be assigned to components from the analyzed samples (e.g., high frequency noise, low frequency noise, drift)

Overload: the sample mass injected onto the column at which efficiency and resolution begins to be effected if the sample size is increased further

Pascal (Pa): a unit of pressure. 1 MPa = 10 bar (atm) = 145 psi = 9.87 atm

Particulate matter: small solid particles in the sample or mobile phase, which can plug the column or other parts of the chromatographic system

Peak (chromatographic peak): a plot of the detector's response in a chromatogram caused by a certain property of an analyte as it passes through the detector's flow cell

Peak width: a measurement accomplished either at the peak-base level or at its half height

- **the baseline width** is the distance between the points where the tangents drawn to the inflection points on the front and back slopes of the peaks intersect the baseline, given in units of time.
- **the width at half height** is the distance from the front slope of the peak to back slope, measured at 50% of the maximum peak height.

Photodiode array detector (PDA): a spectrophotometric detector that allows the simultaneous detection at multiple wavelengths, usually in the range of 190–900 nm

Plate number (N): a measure of efficiency which describes how good a column is in keeping peaks narrow. Columns with large plate numbers give narrow peaks; long columns packed with small particles give the highest plate numbers.

Polystyrene–divinylbenzene (PS-DVB): a common solid polymeric packing used in ion-exchange chromatography, in which ionic groups are incorporated by various chemical reactions

Precision: the extent to which the results of replicate analyses using a given method are close to each other, so that repetitive analysis of the same sample gives similar results

Precolumn: a small column placed between the pump and the injector intended to pre-condition the mobile phase

Preparative chromatography: a branch of chromatography used to isolate and/or to purify components from complex mixtures which involves high amounts of samples, large columns and high amounts of mobile phases. The corresponding systems use fraction collectors coupled with the chromatographic system, to separate the component(s) of interest.

Recovery: the percentage of a target analyte extracted from a sample having a known concentration of the analyte, usually measured as the ratio of the response of the system to an extracted sample; a good recovery is important for adequate quantitation

Regeneration: a procedure intended to restore the stationary phase to its initial state

Retention time: (t_R) the time between sample injection and the peak apex; when all chromatographic conditions are held constant, the retention time for a given peak remains constant

Robustness: the ability of the method to remain unaffected by variations of method and environmental parameters (e.g., dilution of the sample, pH, temperature and sample storage conditions); robustness has to be established under conditions in which the aforementioned parameters are varied in a controlled fashion

Selectivity: the ability of a chromatographic column to separate different analytes

Sensitivity: the ability of a detector to give larger peaks and a better signal-to-noise ratio, hence providing more precise analysis of very small sample concentrations

Solid-phase extraction (SPE): a technique for sample preparation using a solid-phase packing

Specificity: the ability of the analytical method to establish the presence of the analyte(s) of interest in the presence of other compounds that may be expected to be present in at least some samples

Standard operation procedure: procedures describing in detail the analytical steps for determination of certain analyte(s), meant for ensuring that a high rigor is applied for both analysis and data processing

Standard solution: a solution of pure compound(s) with accurately known concentration(s), that is to be determined by a chromatographic procedure

Strong anion exchanger: anion-exchange packing with strongly basic ionogenic groups (e.g., tetraalkylammonium groups)

Strong cation exchanger: cation-exchange packing with strongly acidic ionogenic groups (e.g., sulfonate groups).

Suppressor: a device placed after the ion-exchange column for removing or suppressing the ionization of certain ions so that sample ions can be observed in a weakly conducting background by using a conductivity detector.

Tailing: the phenomenon in which a chromatographic peak presents an extended trailing edge, having an asymmetry factor greater than 1; it can result from injecting an excessive mass or sample, badly packed columns, excessive extra column volume, poor fittings

Tailing factor: a measure of peak asymmetry defined as the ratio of the peak width at 5% of the apex to the distance from the apex to the 5% height on the short time side of the peak

Test chromatogram: chromatogram supplied by a column manufacturer that usually accompanies a new column, showing the separation of certain compounds under a set of experimental conditions

Troubleshooting: locating the cause of problems encountered while operating certain equipment, then solving these problems

Validation: a planned experimental procedure in which the performance of an analytical method is established by the measurement of samples having known characteristics, in order to demonstrate that this performance suffices for the application for which it is proposed

Void: a space or gap formed usually at the head of a column, caused by a settling or dissolution of the column packing, which can lead to decreased efficiency and loss of resolution

Void time: (dead time, holdup time): the retention time of the first, unretained peak

Weak anion exchanger: a stationary phase with weakly basic ionogenic groups (e.g., amino diethylaminoethyl groups)

Weak cation exchanger: a stationary phase with weakly acidic ionogenic groups (e.g., carboxyl groups)

References

Armbruster, D. A., & Pry, T. (2008). Limit of blank, limit of detection and limit of quantitation. The Clinical Biochemist Reviews, 29, S49–S52.

Berek, D. (2010). Size exclusion chromatography–a blessing and a curse of science and technology of synthetic polymers. Journal of Separation Science, 33(3), 315–335.

Blumberg, L. M., & Klee, M. S. (2001). Metrics of separation in chromatography. Journal of Chromatography. A, 933(1–2), 1–11.

Braithwaite, A. & Smith, J. F. (2012). Chromatographic methods. Springer Science & Business Media Dordecht.

Corradini, D., Eksteen, E., Eksteen, R., Schoenmakers, P., & Miller, N. (1998). Handbook of HPLC. CRC Press, Boca Raton.

D'Amore, T., Di Taranto, A., Berardi, G., Vita, V., & Iammarino, M. (2021). Going green in food analysis: A rapid and accurate method for the determination of sorbic acid and benzoic acid in foods by capillary ion chromatography with conductivity detection. LWT, 141, 110841.

Dong, M. W. (2006). Modern HPLC for practicing scientists. John Wiley & Sons Inc. Hoboken, New Jersey.

Frenzel, W. (2002). Dialysis as an automatable sample preparation technique for ion chromatography. In Sample preparation techniques for ion chromatography. Metrohm Ltd., Herisau, Switzerland, 61–74.

Fritz, J. S., & Gjerde, D. T. (2009). Ion chromatography, 4th ed. Wiley-VCH Verlag GmbH & Co. KGaA, Weinheim.

Fritz, J. S. (1987). Ion chromatography. Analytical Chemistry, 59(4), 335A–344A.

Haddad, P. R., & Jackson, P. E., (1990). Ion chromatography. Elsevier, Amsterdam.

Haddad, P. R., Jackson, P. E., & Greenway, G. M. (1991). Ion chromatography: Principles and applications. Elsevier, Amsterdam.

Hewavitharana, A. K. (2009). Internal standard-friend or foe?. Critical Reviews in Analytical Chemistry, 39(4), 272–275.

Jackson, P. E., Romano, J. P., & Wildman, B. J. (1995). Studies on system performance and sensitivity in ion chromatography. Journal of Chromatography. A, 706(1–2), 3–12.

Liu, J. M., Liu, C. C., Fang, G. Z., & Wang, S. (2015). Advanced analytical methods and sample preparation for ion chromatography techniques. RSC Advances, 5(72), 58713–58726.

Majors, R. E., & Carr, P. W. (2001). Glossary of liquid-phase separation terms. LC GC: Magazine of Liquid and Gas Chromatography, 19(2), 124–163.

Majors, R. E., & Hinshaw, J. (2013). The chromatography and sample preparation terminology guide. LC-GC North America, 31(10), S43–S43.

Martin, A. J., & Synge, R. L. (1941). A new form of chromatogram employing two liquid phases: A theory of chromatography. 2. Application to the micro-determination of the higher monoaminoacids in proteins. Biochemical Journal, 35(12), 1358.

McGorrin, R. J. (2009). One hundred years of progress in food analysis. Journal of Agricultural and Food Chemistry, 57(18), 8076–8088.

Meyer, V. R. (2013). Practical high-performance liquid chromatography. John Wiley & Sons, Chichester.

Michalski, R. (2016). Principles and applications of ion chromatography. In Michalski, R. (ed.), Application of IC-MS and IC-ICP-MS in environmental research. John Wiley & Sons Inc, Hoboken, New Jersey, 1–46.

Michalski, R., & Pecyna-Utylska, P. (2020). Ion chromatography as a part of green analytical chemistry. Archives of Environmental Protection, 46(4), 3–9.

Michalski, R., & Pecyna-Utylska, P. (2021). Green aspects of ion chromatography versus other methods in the analysis of common inorganic ions. Separations, 8(12), 235–244.

Mori, S., & Barth, H. G. (1999). Size exclusion chromatography. Springer Publisher Berlin-Heidelberg.

Paull, B., & Nesterenko, P. N. (2013). Ion chromatography: Fundamentals and instrumentation. In S. Fanali, P. R. Haddad, C. F. Poole, P. Schoenmakers, & D. Lloyd (Eds.), Liquid Chromatography. Elsevier, Amsterdam Vols. 157–191.

Robards, K., & Ryan, D. (2021). Principles and practice of modern chromatographic methods. Academic Press, New York.

Saubert, A., Frenzel, W., Schafer, H., Bogenschutz, G., & Schafer, J. (2004). Sample preparation techniques for ion chromatography. Metrohm, Herisau, Switzerland.

Settle, F. A. (2002). Analytical chemistry and the Manhattan project. Analytical Chemistry, 74, 36A–43A.

Snyder, L. R., Kirkland, J. J., & Dolan, J. W. (2011). Introduction to modern liquid chromatography. John Wiley & Sons Inc, Hoboken, New Jersey.

Steinbach, A., & Wille, A. (2009). Determining carbohydrates in essential and non-essential foodstuffs using ion chromatography. LC-GC Europe, 3(7), 15–18.

Tswett, M.. (1906). Adsorptionsanalyse und chromatographische methode. Anwendung auf die chemie des chlorophylls. Berichte der Deutschen Botanischen Gesellschaft, 24, 384–393.

Tsvett, M.. (1903). On a new category of adsorbance happenings and its appliance to biochemical analysis. Trudy Varshavskogo Obschestva Yestestvoispytateley. Proceedings of the Warsaw Naturalists Society, 14(6), 1–20.

Tswett, M. (1901). Paper chromatography. Travaux de la Societe des Naturalistes de Kazan, 35, 268.

Wang, Y. (2010). Determination of nitrite and nitrate in fruit juice by in-line dialysis-ion chromatography. Chinese Journal of Health Laboratory Technology, 12, 058–065.

Weiss, J. (2016). Handbook of ion chromatography. John Wiley & Sons Hoboken, New Jersey.

Xiong, Z., Dong, Y., Zhou, H., Wang, H., & Zhao, Y. (2014). Simultaneous determination of 16 organic acids in food by online enrichment ion chromatography – mass spectrometry. Food Analytical Methods, 7(9), 1908–1916.

Xu, J., Lei, M., Zhu, Z., Yu, Z., Li, D., & Yu, Q.. (2014) Determination of polyphosphates in caviar using ion chromatography-inline ultrafiltration. Journal of Food Safety and Quality, 5, 2753–2756.

2 Ion chromatography

Among liquid chromatographic techniques, ion chromatography (IC) occupies a distinct place, gained mainly in the last few decades; it has been reviewed by several authors (Dasgupta, 1992; Michalski, 2014a) and is covered extensively in numerous books (Foster, 2005; Haddad and Jackson, 1990; Fritz and Gjerde, 2010; Weiss, 2016).

The IC technique, introduced by Small et al. (1975) as a new analytical method, was originally developed to allow the use of conductivity detection at high sensitivity. Since then, IC progressed from a new analytical alternative for a few inorganic anions and cations to a powerful technique for many ionic or ionizable species, including organic ones; the development of separator columns with high efficiencies resulted in a significant reduction of analysis time, while the newly designed electrochemical and spectroscopic detectors further enlarged the area of applications, with improved sensitivity (Michalski, 2014). As a result, IC is a preferred analytical method in many fields, such as environmental monitoring, biochemical research, clinical analysis, pharmaceutical companies, chemical and petrochemical companies, electronics' manufacturers, food industry, power plants, water providers, wastewater treatment facilities, mining/metal/plating companies and universities. Detection limits are typically in the low mg/L range for most inorganic analytes under standard operating conditions, although they can be significantly lower, depending upon the application (Jackson et al., 2001; Michalski, 2014; Parab et al., 2021.). IC is based on different separation mechanisms: ion-exchange, ion-exclusion and ion-pair chromatography. Depending on the separation mode, different types of stationary phases are used: for example, for ion exchange, the stationary phase is either an anion or cation exchanger, while for ion-pair separation, a reversed-phase stationary phase is used (Haddad and Jackson, 1990).

Ion-exchange chromatography (IEC) is the most relevant for IC, since the majority of IC applications are established on the IEC mechanism. It is based on an equilibrium process occurring between the ions from the mobile phase and ion-exchange groups bonded to the stationary phase; this equilibrium is dependent on the temperature, pH, nature and concentration of the present counter ions. IEC is applied for the separation of both organic and inorganic anions and cations. The separation of ionic analytes takes place on the basis of ion exchange in stationary phases with charged functional groups. Common stationary phases used in IEC are based on polystyrene (PS) ethylvinylbenzene, or methacrylate resins copolymerized with divinylbenzene (DVB), bearing different functional groups for cation or anion exchange; the functional groups, typically, are quaternary ammonium groups for anion exchange and negatively charged groups like sulfonates or carboxyl for cation exchange. The mobile phase usually contains an ion with a charge opposite to the ionic group on the surface of the stationary phase and in equilibrium with this by forming an ionic couple (Acikara, 2013; Cummins et al., 2017).

https://doi.org/10.1515/9783110644401-002

Most of IC separations occur by ion exchange in stationary phases with charged functional groups (Dorfner, 1991; Fritz and Gjerde, 2009); the corresponding counter ions are located in the vicinity of the functional groups and can be exchanged with other ions having similar charge from the mobile phase; hence, various ionic components of the sample can be separated based on their differential affinities towards the immobilized stationary and the liquid mobile phase, as follows:

– for cation exchangers:

$$R - SO_3^- H^+ + Na^+ \rightleftharpoons R - SO_3^- Na^+ + H^+$$

– for anion exchangers:

$$R - [N^+ (CH_3)_3]OH^- + Cl^- \rightleftharpoons R - [N^+ (CH_3)_3]Cl^- + HO^-$$

For each ionic species, the exchange process is characterized by a corresponding equilibrium, which determines the distribution between the two phases. Different ionic species from mixtures can thus be separated on the basis of their different affinities for the stationary phase (Weiss, 2016).

The ion-exchange capacity of a resin can be expressed as the number of ion-exchange sites per weight equivalent of column packing and is typically expressed in µEq/g resin or µmol/g. In IC, only low-capacity exchangers are used, their exchange capacity showing important changes with pH.

During the separation of ions by ion-exchange columns, small ions are eluted before larger ions and monovalent ions before di- and tri-valent ions; for example.:

– the separation sequence for anions: $F^- > Cl^- > NO_2^- > NO_3^- > PO_4^- > SO_4^-$;
– the separation sequence for cations: $Li^+ > Na^+ > K^+ > Ca^{2+} > Mg^{2+}$.

Unlike inorganic ions, which are small and densely charged structures, organic species are much larger, with a lower charge density; hence, it is necessary to work with both weak and strong anion- and cation-exchange columns containing the following types of bonded phases (Acikara, 2013):

– strong anion exchangers (SAX), for example, quaternary ammonium groups;
– weak anion exchangers (WAX), for example, tertiary amino groups, usually diethylaminoethyl (DEAE);
– strong cation exchangers (SCX), for example, sulfonic acid, usually on a propyl chain;
– weak cation exchangers (WCX), for example, with carboxylic acid, usually carboxymethyl.

The stationary phases used in IEC are made of small particles (2–10 µm) with spherical shapes, which differ in the type of the support material, in their pore sizes and in their ion-exchange capacities. Stationary phases based on silica gel were the first used in IC; they had good separating performances and very good mechanical properties, but showed a relatively low chemical stability, which limited their use for a narrow pH range (~2–7). Later, ion exchangers based on different polymers became more popular, since they exhibit greater chemical stability. Styrene-divinylbenzene copolymers, polymethacrylate and polyvinyl resins, as well silica gels with chemically bounded phases are the most used materials in the manufacturing process for ion exchangers (Michalski, 2009; Weiss and Jensen, 2003).

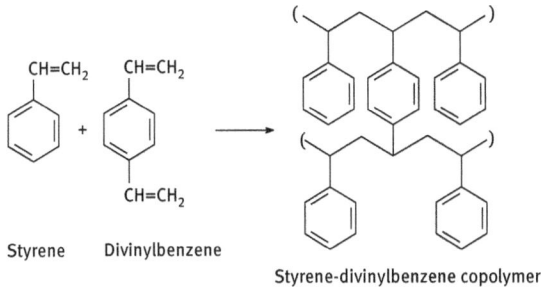

Figure 17: The copolymerization of styrene and divinylbenzene.

Among these, styrene-divinylbenzene copolymers are frequently employed as substrate materials (Figure 17), and cation exchangers can be obtained by sulfonation of this resin, while anion exchangers can be obtained by chloromethylation followed by amination. The most important functional groups in anion chromatography are derived from trimethylamine and dimethylethanolamine (Figure 18), while for cation chromatography, the majority of commercial cation exchangers use sulfonic acid groups.

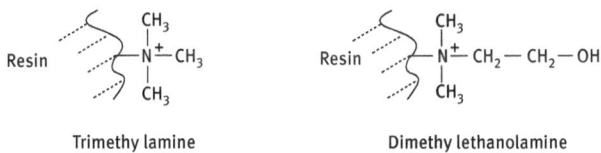

Figure 18: The most common functional groups used in column packings for anion chromatography.

Ion-exclusion chromatography involves steric exclusion, sorption processes and even hydrogen bonding. The separations are accomplished using high capacity totally sulfonated PS-DVB-based cation-exchange stationary phases. Ion-exclusion chromatography is particularly useful for the separation of weak inorganic and organic acids; acids with high strengths are not retained and elute unseparated within the void volume (Tanaka and Haddad, 2000).

Ion-pair chromatography is based mainly on adsorption, analytes being separated after pairing with a surface-active counter ion; traditionally, C_{18} stationary phases used in reversed-phase HPLC can be used, also in this case. The mobile phase has to contain an appropriate ion-pairing reagent, taking into account the chemical nature of the analytes. This mechanism is suited for the separation of surface-active anions and cations as well as for transition metal complexes (Cecchi, 2009).

Overall, IC is a liquid chromatography technique that can separate solute mixtures based on the charge and the magnitude of the ionic species from the sample, as a result of interactions with a stationary phase with charged functional groups. It is used for the separation of ionic species from mixtures, for removing interfering substances from a sample or for separation of larger amounts of compounds in semi-preparative or preparative applications.

In food analysis, IC is used mainly to know *what* analytes are present in a sample and/or *how much* of each analyte is present, in other words, for identification and/or for the quantification of the components from a sample. The corresponding techniques for these purposes are qualitative analysis (for establishing the identity of different compounds in the sample) and quantitative analysis (for measuring the concentration or amount of each analyte). Usually, IC deals with quantitative analysis, but this involves, as a preliminary stage, the identification of the compounds whose concentrations one measures.

2.1 Instrumental setup

IC systems ensure appropriate conditions for analytical, preparative or process applications, in order to achieve fast and reproducible separations using high pressure, controlled temperature and in-line detection. The instrumentation used for IC is similar to that employed in conventional high-performance liquid chromatography (HPLC); in fact, they can be considered HPLC systems designed for the analysis of ionic compounds, having wetted surfaces typically made of an inert polymer, such as polytetrafluoroethylene (PTFE), or, more commonly, polyetheretherketone (PEEK), rather than stainless steel (Haddad et al., 1991). This is due to the fact that the corrosive eluents and regenerant solutions used in IC can contribute to corrosion of stainless steel components.

Typical components include a pumping system, a sample injector, a column, an optional suppressor, a flow-through detector and a chromatographic data system, as shown in the schematic diagram given in Figure 19 (Ahuja et al., 2021; Weiss, 2016; Wouters et al., 2017). Conventional HPLC systems can be configured for performing certain IC analyses, provided that their components are compatible with the mobile phase and the sample.

The mobile phase is supplied from one or several reservoirs; devices that electrolytically generate acid or base eluents are also commercially available. An in-line vacuum degasser unit removes the dissolved gasses; then, a pumping system delivers the mobile phase through the system. The most widely used are dual-piston pumps (mechanical reciprocating pumps with pistons operating out of phase to provide a constant flow rate). Any part of the system that is in contact with the mobile phase should be constructed of materials inert to corrosive components.

The sample is injected into the IC system through an injection module; after switching the injection valve, the sample is transferred in the flow of the mobile phase. Typical injection volumes are in the range of 10–100 μL. Manual injection is the cheapest alternative; it uses small-volume syringes to fill a fixed-volume loop attached to the injection valve. To achieve higher reproducibility, automatic injectors are the best choice, these enabling also a higher productivity. Besides automatically injecting samples, modern autosamplers can perform dilution and automated sample preparation, as well.

The heart of all IC systems is the separator column; it contains the stationary phase packed in a tube manufactured from an inert material. The choice of a suitable stationary phase and chromatographic conditions determine the quality of the analysis. Since maintaining a certain temperature is important for most separations, the column is hosted in a thermostated oven.

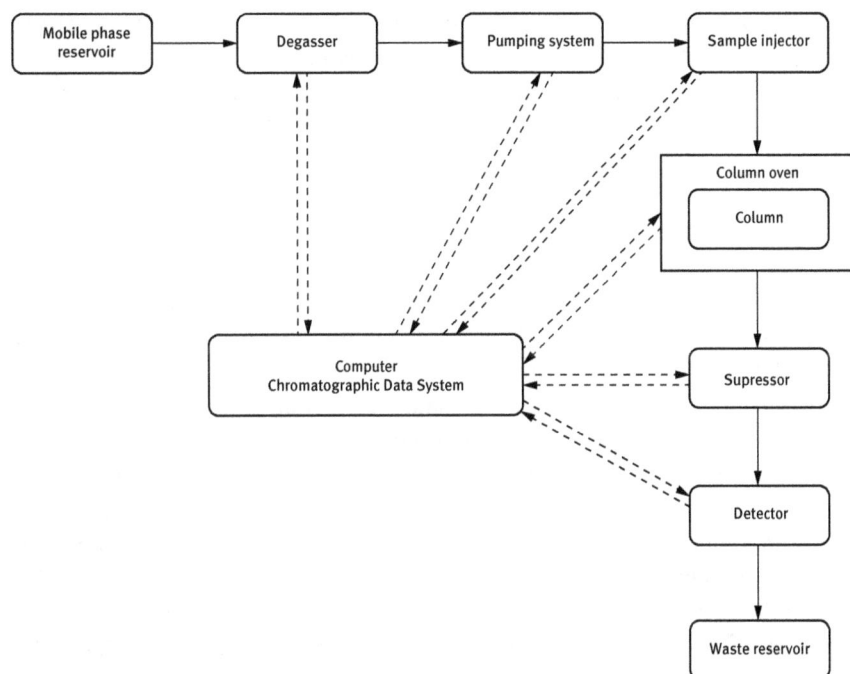

Figure 19: Block scheme of an ion chromatograph.

The suppressor is a module that can reduce high background conductivity of the electrolytes in the eluent, while enhancing the conductivity of the analytes; it converts certain components from the eluent into low conducting species, reducing the signal from the eluent species and simultaneously enhancing the signal of the analyte, thus, improving the signal-to-noise ratio of the detection system.

The detector monitors the eluate as it leaves the column, generating an electronic signal proportional to the concentration of each separated component. The most commonly employed detector in IC is the conductivity detector. Other detection methods, such as ultraviolet/visible (UV/VIS) absorption or amperometry, have proven to be highly sensitive for certain absorbing or electroactive species, while post-column derivatization followed by UV/VIS detection is an important approach for the detection of particular ions. The use of advanced detection techniques, such as mass spectrometry (MS) and inductively coupled plasma (ICP) spectrometry, coupled to IC separations continues to increase; the obvious disadvantage of these detection techniques is that they each add considerable complexity and significant cost to the analysis.

The computer is a must, since it controls the whole chromatographic system; it has to run a specialized software (chromatographic data system), which provides the user interface, diagnoses the system, controls the separation parameters, accomplishes data acquisition, as well as identification/quantification/quality control procedures and, finally, report generation and data storage.

The next sections provide further details on each module for a better understanding of the IC systems' functions.

2.1.1 The mobile phase

The mobile phase is located usually at the top of the IC system, in glass or plastic reservoir(s); an inlet frit or filter made of glass or polyethylene, with 5–10 μm pores, is fixed at the entrance point at the Teflon tubing, which connects the reservoir with the degasser, to keep possible particulate matter from entering the IC system, while holding the inlet line at the bottom of the reservoir (Figure 20). The reservoir should have graduations marked to help the user estimate the available volume and a proper cap to prevent particulate matter from contaminating the mobile phase. The attachment of an adsorption tube to the reservoir's cap can prevent carbon dioxide contaminating the mobile phase.

Despite the relatively large pore size, the inlet frit (also called sinker frit) can get plugged over time, especially as a result of microbial growth; hence, it has to be tested periodically and changed, if the mobile phase does not flow freely. Flushing with isopropylic alcohol can prevent microbial growth.

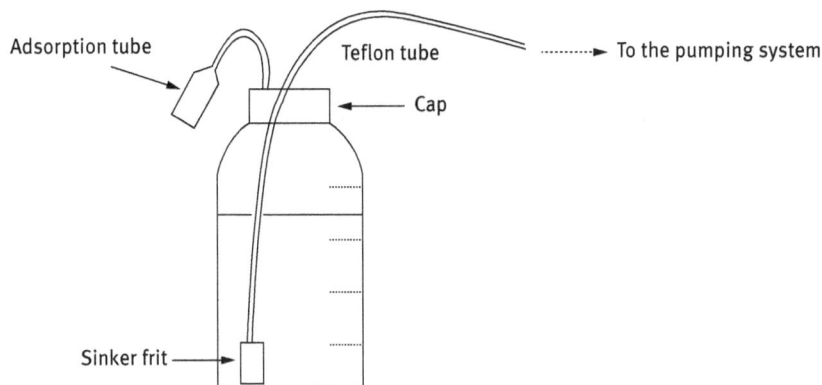

Figure 20: A schematic of the mobile phase reservoir.

The mobile phase has to be compatible with the detector and must not interact chemically with the system; otherwise, corrosion phenomena, salt precipitations, etc. can occur, damaging the system.

A good-quality mobile phase is essential for IC separations; its composition is directly related with the column and the detector used for a certain analytical issue. Common mobile phases in IC are generally dilute aqueous solutions of acids, alkalis or salts such as HNO_3, methanesulfonic acid, phthalic acid, p-hydroxybenzoic acid, KOH, NaOH, $NaHCO_3$ and Na_2CO_3. For separations of anions, salts of phthalic or p-hydroxybenzoic acids (Figure 21) are commonly used for direct IC, since these have a low conductivity combined with a high elution power. For cations of alkali and alkaline earth metals, IC is generally performed with diluted acids such as nitric acid or methanesulfonic acid. The concentration range for the mobile phases used in IC is very low (several mmol/L).

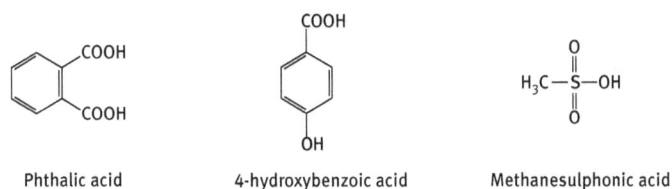

| Phthalic acid | 4-hydroxybenzoic acid | Methanesulphonic acid |

Figure 21: Structures of several acids used for the preparation of mobile phases.

As conductivity detectors are commonly used in IC, the mobile phases have to be pre-pared with high purity water (ultrapure water) and high purity substances (pro-analysis); type I water is ideal (18.2 MΩ, 0.056 μS), providing better sensitivity and a more stable baseline. A laboratory water purification system producing high purity water is necessary, in order to provide the necessary volumes for preparation of mobile phases, rinsing, dilutions, etc. Multicomponent mobile phases are prepared by measuring the required volumes and weighing the necessary masses, followed by mixing. Adjustment of the pH, if necessary, is accomplished using aqueous solutions on an efficient stirrer and monitoring the pH with a properly calibrated instrument.

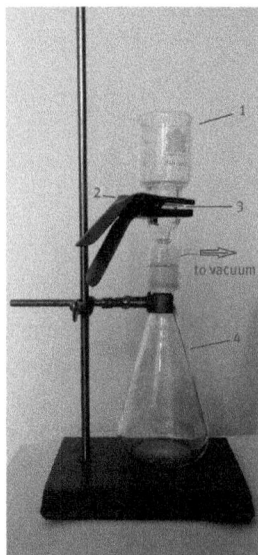

Figure 22: Vacuum filtration system: 1, feeding funnel; 2, clamp; 3, membrane filter; 4, collector flask.

To prevent bacterial growth, the mobile phase should be made freshly and not used for a long period of time; 5% methanol or acetone can be added, as a preventive measure.

As small changes in the mobile phase composition can cause important differences in retention times, making an accurate mobile phase is crucial in chromatographic analysis; hence, the following rules have to be followed:
- All glassware and utensils used in mobile phase preparation must be clean (any impurity may dissolve and contaminate the mobile phase).
- High purity water and reagents must be used.
- Volume measurement and weighing has to be performed with accuracy.
- When preparing solutions, the mixture should be thoroughly shaken, allowing the air bubbles to be released before attempting to measure the volume.
- The mobile phase has to be filtered using vacuum filtration through membrane filters to remove particles greater than 0.45 µm (Figure 22), then degassed (e.g., by applying vacuum from a water-jet or a vacuum pump for 5–10 min, while keeping the bottle on an ultrasonic bath), before the transfer in the mobile-phase reservoir.
- When refreshing the mobile phase, it is not recommended to pour the freshly prepared mobile phase into old one; it is better to remove the reservoir from the

system, to empty the old mobile phase, to rinse the reservoir with the newly prepared one and fill the reservoir with the fresh mobile phase.
– Certain mobile phases (such those containing sodium hydroxide) can absorb carbon dioxide from the ambient air, resulting in composition changes; hence, supplementary precautions are necessary to avoid this (by using CO_2 trap absorber cartridges mounted on the mobile-phase reservoirs' caps).

Most of the commercially available IC systems require the manual preparation of mobile phases, but some of them are available with electrolytically generated mobile phases (a dedicated module produces the necessary mobile phase by means of a rigorously controlled electrical current), this case being discussed later (Section 4.1).

2.1.2 The mobile-phase degasser

Degassing is an important stage during mobile-phase preparation, intended to eliminate the dissolved gasses; it is recommended to perform degassing immediately after the preparation of the mobile phase by applying vacuum, while the mobile phase is kept on an ultrasonic bath. In an IC system, it can be accomplished either by using a helium stream directed in the mobile-phase reservoir ("helium sparging"), or mostly by using an in-line vacuum module (Dolan, 2014b); despite helium sparging being the most effective degassing option, it is not a convenient one for most laboratories (supplementary technical problems and higher operational cost).

Degassing is a must, since the dissolved gasses may come out of the mobile phase as bubbles in the pumping system, in the column or in the detector cell and cause baseline spikes and noise.
– In the pumping system, gasses can cause flow rates fluctuations (cyclic pressure fluctuations are indicators of such a problem) or can even cause pumps to lose their prime.
– In the chromatographic column, gasses can disturb the separations, leading to broader peaks.
– If gas bubbles reach the detector flow cell, they can cause noise.

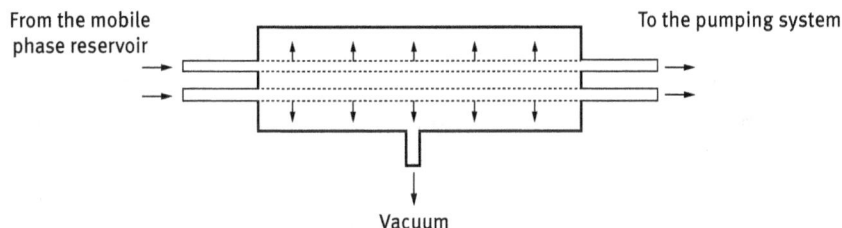

Figure 23: Simplified scheme of an in-line degasser.

In-line vacuum degassers use semipermeable membranes, which allow the migration of dissolved gases from the mobile phase in a vacuum chamber; hence, gases dissolved in the mobile phase diffuse through the membrane, then, the degassed mobile phase is directed to the pumping system (Figure 23).

2.1.3 The pumping system

A relatively high pressure is required to force the mobile phase through the chromatographic column, which is packed with small particles that oppose high resistance to the flow. A constant flow has to be delivered, regardless of the pressure caused by the flow resistance of the column (back pressure); this is a must for reproducible retention in analysis, since the common detectors are mass-sensitive (their output is related to the flow rate).

The pumping system has to deliver a metered constant flow rate of the mobile phase (from <1 to 10 mL/min), at a high pressure (~ 350 bar), with minimal fluctuations over extended periods of time, from the mobile-phase reservoirs to the column; then, to the detection module through connecting tubing and fittings. Constant-flow pumps allow pressure to vary, while maintaining a constant flow rate.

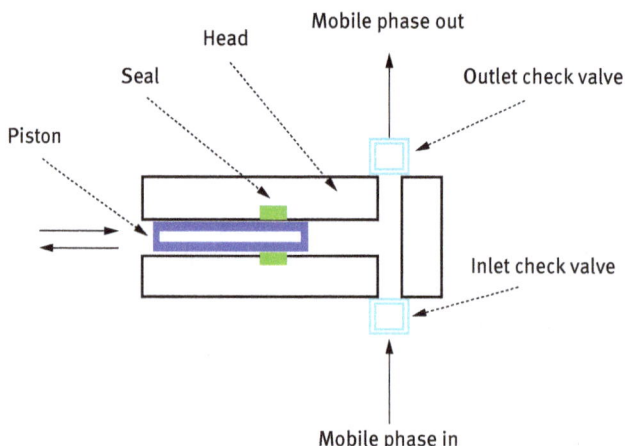

Figure 24: Schematic diagram of a reciprocating-action pump.

Common pumps used in IC systems are single-head reciprocating-action pumps (Figure 24), in which a motor-driven cam pulls a piston back and forth in the pump head; a pair of check valves serves to maintain a one-way flow of solvent – one opens the pump head chamber to the mobile-phase reservoir during the piston's fill stroke and closes the path to the mobile-phase reservoir during the pump delivery stroke; the other one works in the reverse, while a special seal placed around the piston prevents leakage of the mobile phase out of the pump. The mobile phase is, hence, drawn into the chamber by suction and forced into the column under pressure at each pair of movements of the piston: during the delivery stroke, flow increases from zero up to a maximum, then decreases back, while during the intake stroke, the flow is zero. The pump heads can be made of 316 stainless steel or of PEEK, the latter being preferable for IC. The pumping system includes a purge valve, used to help in priming and in removing gas bubbles, which are trapped occasionally (LaCourse and LaCourse, 2017; Meyer, 2013).

The pressure changes in a similar manner as the flow causing pressure pulses; the discontinuous pump cycle, which causes pulsations in pressure and flow rate, is an important disadvantage of this type of pump. Such pulsations can be dampened by using different technical solutions, such as: using special cam shapes, varying the speed of the motor, using a pulse dampener, or using pumps with two heads, which oppose each other in phase, such that the intake stroke from one head coincides with the delivery stroke from the second head. Dual-head reciprocating pumps have lower mobile-phase pulsation but are more complex. Alternatively, pumps controlled by proportioning valves, which mix the mobile phases in the desired proportion, may be used (LaCourse and LaCourse, 2017; Shoykhet et al., 2019; Snyder et al., 2011).

During the last few years, renewed interest in micro-separations has occurred, syringe pumps being good alternatives. In syringe pumps, the mobile phase is displaced by the action of a stepping motor, being delivered to the column at the defined flow rate. An obvious disadvantage of this type of pump is the limited volume of the syringe, which requires periodic refilling; however, when using small-diameter columns, these pumps can be the best choice, since they provide very stable flow rates. These pumps can deliver pulsation-free flow rates, being ideal for flow-sensitive detectors (e.g., electrochemical or mass spectrometric detectors). Syringe pumps may be provided with a purging device for priming purposes and for removing the entrapped air (Shoykhet et al., 2019).

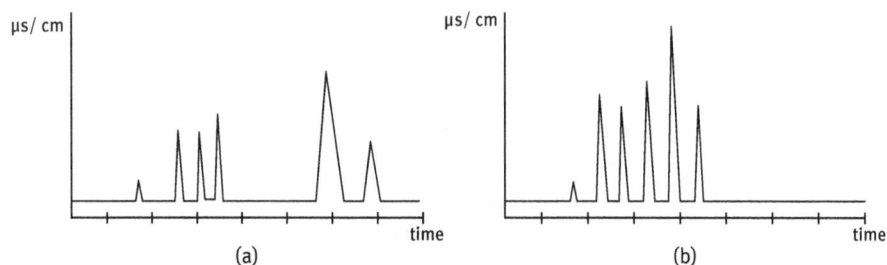

Figure 25: Isocratic (a) versus gradient (b) separations.

If the composition of the mobile phase is constant throughout a separation, the elution is called **isocratic** (without changes in the composition of mobile phases). For samples containing complex mixtures of analytes with high differences in properties, an isocratic elution can lead to very long separations in which the analytes with small k′ values are not well separated, while compounds with large k′ values need excessive separation times. In such cases, it is more practical to gradually change the composition of the mobile phase, thus changing the retention of late-eluting compounds, by using gradient elution. **Gradient elution** involves a continuous change in mobile-phase composition; this provides faster separations (Figure 25), with improved limits of detection and less tailing, for most analytes. Gradient elution can also overcome the problem caused by certain compounds with a high affinity for the stationary phase that they do not elute at all (Snyder, 1980).

When a change in the composition of the mobile phase is necessary during the separation of complex mixtures, either two pumps or pumps controlled by proportioning valves are a must, in order to achieve a gradient elution. The proportions of the mobile phases involved are controlled by the chromatographic data system.

There are two common methods for mobile-phase mixing and gradient formation (Figure 26):
- **Low-pressure mixing** – uses a single high-pressure pump and a proportioning valve providing the desired ratio of mobile phases, which are mixed on the upstream side of the pump.
- **High-pressure mixing** – uses a separate pump for each mobile phase; then, the pump outputs are combined at high pressure.

To resume, in the pumping system the mobile phase gets pumped and pressurized (in order to be able to pass through the column), dampened (to have a pulsation-free flow) and eventually proportioned and mixed (when working with gradient elution).

Irrespective of the elution type, the pumping system has to be operated with care, in order to avoid trouble and to prolong the pumps' lifetime; it must not be allowed to run dry, and aqueous buffers must not be left in the pumps when these are not in operation. **In-line filters** (see #4.1.2) placed between the mobile phase reservoir and the pumping system are recommended to prevent solid particles from entering the pumps. The pumping system cannot tolerate pressures above a certain limit, which is usually 250–400 bar or 25–40 MPa; most pumps allow the setting of an upper and lower pressure limit, such that when pressure falls outside these limits, the motor stops. The upper pressure limit is for the protection of the equipment and should be set as low as possible; when using PEEK connections, it will need to be around 3000 psi. The lower pressure limit is used to prevent the system running if the solvent reservoir runs dry, or a major leak occurs or large amount of gases get trapped in the pump; such a limit is set usually when the system is intended to run unsupervised, using a value that is at least 500 psi lower than the normal operating pressure for the system – the higher the pressure limit, the more sensitive the system will be and the more likely to cut out (Dong, 2006; McMaster, 2007).

Preventive maintenance of the pumping system can ensure their reliable and trouble-free operation and has to include regular flushing of the system with buffer-

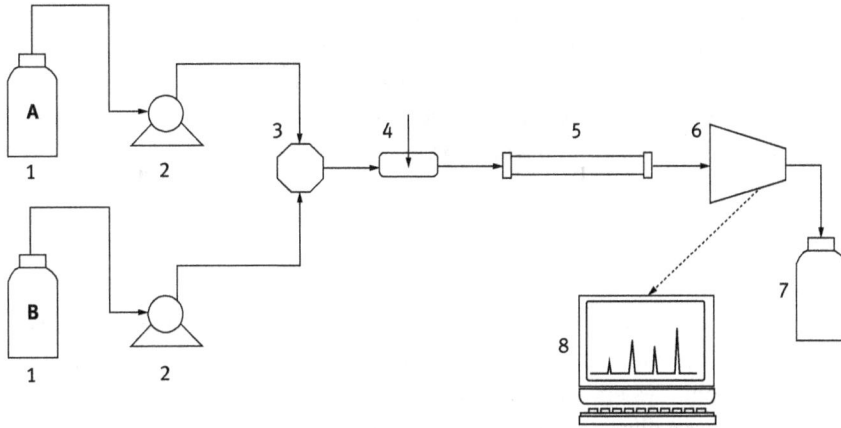

a. high pressure mixing

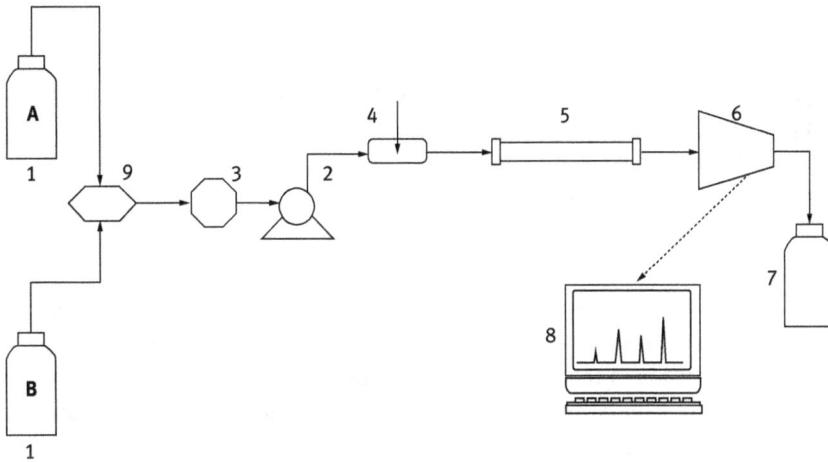

b. low pressure mixing

Figure 26: High pressure versus low pressure mixing. 1, mobile phases reservoirs; 2, pumps; 3, mixer; 4, injector; 5, chromatographic column; 6, detector; 7, waste reservoir; 8, computer; 9, proportioning valve.

and salt-free mobile phase, before shutting down to minimize corrosion, the possibility of bacterial growth in the mobile phase, the precipitation of salts in the pump head or check valves and the deposition of salt crystals under and behind the pump seals. Change of seals is a must when they wear and allow mobile phase to leak out the pump head; the wearing of seals is aggravated by crystallization of buffer salts as solvent evaporates, hence, mobile phases that contain salts should never be left in the IC systems – the system should be flushed with a salt-free mobile phase before it is shut down. A seal needs replacing when the mobile phase's flow rate is reduced, giving rise to longer retention times, decrease in backpressure and appearance of a puddle near the IC system. A good practice is to replace the seals on a scheduled basis than to wait for a leak (Dong, 2006; Haidar Ahmad, 2017).

2.1.4 Injectors

An injector is a device designed to introduce a liquid sample into the pressurized mobile phase within the IC system; a reproducible introduction of the samples in the chromatographic system is a critical factor for an accurate quantitative analysis. The injection process plays an important role in chromatographic separations, as it is one of the sources of extra-column variance (Colin et al., 1979); the injectors have to be located as close as possible to the column to minimize the dead volume. Sample injection can be accomplished manually (by a syringe) or automatically (by autosamplers) (Paul et al., 2019).

Manual injectors are simple, inexpensive and require little experience; they are popular in laboratories where the equipment is used on irregular basis. In manual injection, two position (load/inject) valves are used; these are usually six-port devices, which allow the samples to be loaded into an external loop, by using a syringe at atmospheric pressure (Figure 27). These valves consist of a fixed part (stator) and a mobile one (rotor), with three internal passages that connect alternate pairs of external ports; they can switch between the two positions.

The activation of the valve shifts the solvent flow stream such that the sample loop is incorporated into the flow, and the sample is delivered onto the column:

– When the valve is in the "load" position, the pump is connected to the column inlet, and the sample inlet is connected to one end of the sample loop; the other end of the sample loop is connected to the waste port. Using a syringe, the sample is introduced in the needle port, from which it is directed to the sample loop (5–2,000 μL), the excess going to waste (Figure 27a).
– When the valve is turned to the "inject" position, the sample is taken from the sample loop by the mobile-phase stream, being pushed into the column, while the inlet port is connected to waste; during this stage, the syringe has to be refilled with wash solution to clean out the injection port and the sample loop (Figure 27b).

For the next injection, the valve is rotated back to the load position, a new sample is introduced into the sample loop and the entire process is repeated.

The sample loop is a piece of tubing used to precisely measure the volume of the injected sample solution; the larger the sample loop, the lower the detection limit, but, on the other hand, the chance for overloading increases. For optimum

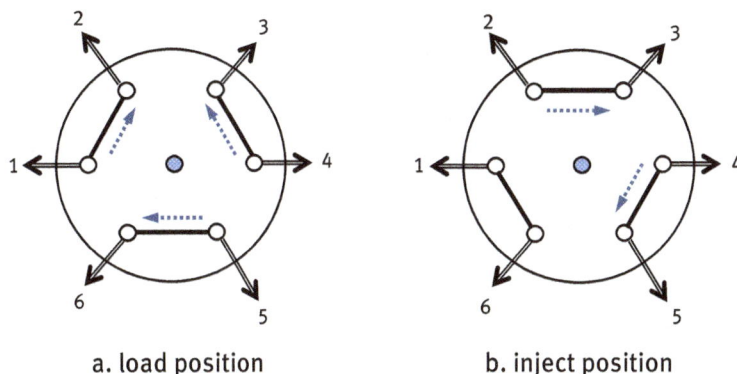

a. load position   b. inject position

Figure 27: Schematic diagram for an injection valve, highlighting the mobile-phase flow. Port's ID: 1, needle; 2, 5, loop; 3, column; 4, pump; 6, waste.

reproducibility, the so-called **filed loop injection** is recommended: this can be accomplished by injecting a 3–4 loop volumes, using any type of syringe fitted to an appropriate flat-ended needle (only one loop volume will be retained, while the rest will go waste but flushing the loop, removing traces of air and ensuring that the loop is completely filled with the sample). If the sample loop is overfilled, syringe error is eliminated, which increases the precision of quantitative analysis, while the accuracy of the injected volume is not dependent on the operator skill. The sample loops may be only partially filled, adding flexibility in terms of injection volume; in such a case, the injected volume is dependent on the operator's experience.

To have a reproducible manual injection, the following procedure has to be followed:

–   The sample has to be loaded in a syringe, avoiding air bubbles (if bubbles are present, they have to be purged).
–   With the injector in the "load" position, the syringe needle is inserted in the sample port, until it engages with the sealing sleeve; then, the syringe piston is pushed slowly until the desired sample volume is introduced, while the excess sample moves out to the waste port.
–   The valve is turned in the "inject position" as fast as possible, to minimize the pressure surge caused by the flow interruption during injection (certain injection valves have a position in which the flow path is completely blocked during the change from "load" to "inject"); while turning the valve is advisable to leave the needle in the injection port, to ensure a tight fit (by withdrawing the needle, a suction effect is generated, causing the content of the waste outlet to be sucked back into the loop).
–   The syringe used to inject samples has to be cleaned after each injection and refilled with wash solution, which has to be injected in the injection port to clean the sample loop.

Autoinjectors (autosamplers) use a mechanically driven version of the six-port valve from manual injectors; nowadays, most IC systems use automatic injection, which is accomplished by autosamplers – computer-controlled injectors that allow unattended operation of the chromatographic systems. In such cases, the sample is introduced by an automated syringe drawing samples from the sample vials, while the injection valve's handle is switched to the necessary position automatically. Autosamplers consist of a carousel or a rack, which holds the sample vials and an injection device to transfer samples from the vials to a loop from which they are loaded into the chromatographic system. In general, autosamplers use the same main parts: needle, sample loop, injection valve, metering device, wash port, needle wash pump, sample rack. The metering device is based on a precision syringe and can handle injection volumes in the range 0.1–100 µL. When using autosamplers, special sample vials have to be used, most of them having 1.5–1.8 mL, being provided with a cap having a pierceable septum (e.g., 1.5 mL polypropylene vials – Thermo Fisher Scientific).

Vial caps for autosamplers are provided with PTFE/rubber/rubber coated with PTFE on one side/rubber coated with PTFE on both sides. PTFE is cheaper, and it does not block the needle, but it cannot reseal; rubber reseals after injections. Rubber with PTFE minimizes the coring problem, where a slug of rubber blocks the needle. For very small amounts of sample, special vials are available.

Autosamplers can be programmed to control numerous issues such as sample volume, the number of injections, loop rinse cycles, the interval between injections, multiple injections from the same sample vial and sample dilution; even the way in which the needle enters the vials can be controlled by the chromatographic data system (needle position, sampling flow rate, etc.). Some producers offer autosamplers with chilling option for preventing the degradation of unstable components from samples, during the time needed to complete a batch of samples.

Autosamplers reduce manual injection steps to a minimum, providing a better reproducibility and accuracy of results, and are advantageous since they:
–   save work because the analyst can perform other tasks while the autosampler is injecting samples;
–   are more reliable, more precise;
–   allow a diverse range of operations (different injection volumes from each vial/ reinjection of a standard from the same vial/duplicate injections of each sample, etc.).

Whatever the injection option, the injected sample volume is an important issue to consider: if this is too small, detection problems can occur (certain analytes will not be detected), while if this is too high, separations can be compromised (overloading, peak tailing, selectivity problems, etc.).

It is important to mention that performing an injection requires an IC system in equilibrium – that means that the mobile-phase flow rate, the pressure and the temperature are constant, the detector is warmed up and provides a flat and stable baseline.

2.1.5 Tubing and fittings

The connections between the IC system's components have to be characterized by a small volume (as small as possible), chemical inertia in relation with the mobile phase and a good resistance to high pressure; tubing and fittings must be made of pressure-resistant material because flow resistance in the column can generate a considerable back-pressure (Dong, 2006; Meyer, 2013).

They are made from Teflon (tubing for low-pressure flows, such as those from mobile-phase(s) reservoir(s) to the degasser and from the degasser to the pumping system) or from specially formulated plastics like poly(ethyl ether ketone) (PEEK) or even from 316 stainless steel (for high-pressure flows). Of these options, stainless steel is the strongest: it can support the weight of a column, it can be bent into different shapes, but, on the other hand, it can be the subject of corrosion in contact with certain mobile phases and may lose metal ions, which can affect chromatography. PEEK is inert, but has a limited capacity to withstand high pressures (up to around 5,000 psi) and needs precautions to go round corners, because it will stretch if bent and then leak. Depending on the internal diameter of PEEK tubing, certain colors are used to easily distinguish between them (Table 2 and Figure 28).

Table 2: Color codes of PEEK tubing.

Color	Internal diameter (mm)	Uses
Natural	0.06	Adding backpressure
Red	0.13	Plumbing 2 mm systems, sample loops
Yellow	0.17	Plumbing 2 mm systems, sample loops
Blue	0.25	Plumbing 4 mm systems, sample loops
Orange	0.50	Waste lines, large sample loops
Green	0.75	Waste lines, large sample loops
Grey	1.00	Waste lines

The fittings used in IC systems are called compression fittings (Figure 29). They have four elements: tubing, a compression nut or screw, one ferrule and a compression body. To connect an end of the tubing with another fitting (the column end, a tee, a connector, etc.) for the first time, a piece of tubing is first inserted into a nut and a ferrule; then, the nut has to be screwed into the fitting and tightened, forcing the ferrule into the fitting body. The ferrule is, hence, compressed onto the outer diameter of the tubing, creating a leak seal. Usually the nut should be tightened fingertight and then, additionally, with a wrench, with care not to overtighten the fitting – not more than to prevent leaks (Walters, 1983). The distance between the ends of the

Figure 28: Colored PEEK tubing.

tubing that the ferrule seats is called **ferrule lock distance**, being an important parameter to consider when accomplishing connections in IC systems.

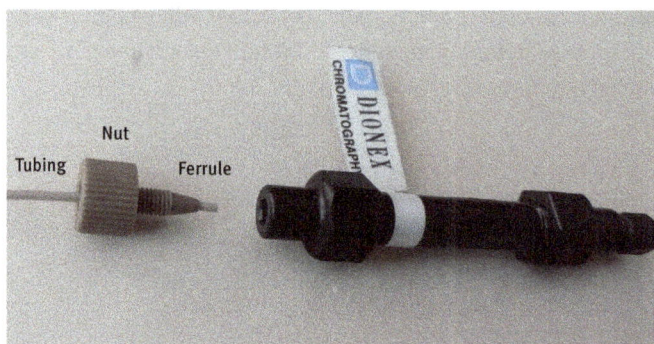

Figure 29: Fingertight compression fitting.

Stainless steel compression fittings are common in HPLC, but they can also be used in IC systems configured for anion analysis; they are advantageous, because they can withstand very high pressures, but they cannot be reused without cutting off and discarding the ferrule, because, in most cases, end-column fittings are different, with different ferrule lock distance:

- If the ferrule lock distance on the tubing is too long for the new fitting, when connecting a column, it may be difficult to engage the screw thread in the end of the column-it will buckle whilst being tightened and may be very difficult to get out again.
- If the ferrule lock distance is too short it will cause a void, causing peak-broadening.

Hence, if steel fittings are to be used it is necessary to cut off the ferrule, to clean the end of the tube, to attach a new ferrule and make the new connection (the ferrule not only locks onto the tubing but also deforms to fit exactly to the internal geometry of the column end-fitting).

Desirable alternatives are the so called Viper fittings (Figure 30), which are designed for "perfect connections" (with no dead-volumes).

Figure 30: Viper fittings.

Ferrules are small pieces of hardware used in chromatography to help prevent leaks; they form a strong, permanent seal around the tubing to fit it with a column, a tee or with an union.

Unions are used to join two pieces of tube fitted with compression screws (Figure 31); some are specially designed to provide a low dead volume, these being recommended when such practice is necessary – otherwise they can introduce additional dead volumes.

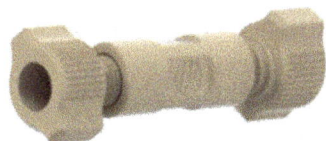

Figure 31: Union assembly.

Compression screws may be made of steel of PEEK; they may be fingertight or hexagon-headed, needing a spanner.

PEEK fingertight compression screws (Figure 32) can be either made of one piece (a cheap option, which seals for low pressures, up to around 3,000 psi, in which the ferrule part turns when the nut is tightened), or two pieces (with separate nuts and ferules, sealing up to around 4,000 psi; they can reuse ferrule, which lasts longer because it does not rotate and can be replaced separately).

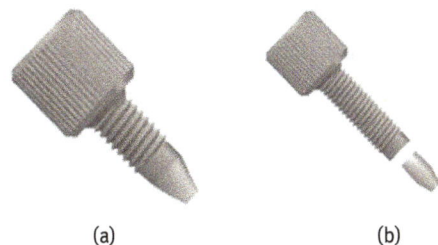

(a)   (b)

Figure 32: One-piece (a) and two-piece (b) fingertight compression screws.

PEEK fingertight compression fittings are advantageous because they can be reused and they do not require a spanner; among disadvantages, one piece fittings wear out much faster and leak at lower back pressure than the two-pieces variety. Once the screw thread or the ferrule gets worn, these must be replaced, or leak will result soon afterwards. When using fingertight fittings, it is important to ensure that the tube is pushed as deep as possible into the fitting, when tightening the nut.

It is important to keep in mind that not all fittings are interchangeable. Although, in some cases, certain fittings appear to work (they join together in the usual way), mismatched fittings can cause separation problems, mainly because of the distance between the tubing end and the ferrule.

- If the distance is too long for the body of the fitting, the ferrule will not seat properly and the fitting will leak.
- If the distance is too short, the ferrule will seat, no leak will occur, but an empty space called "dead volume" will appear.

The abovementioned dead volumes can compromise an otherwise proper separation by band broadening, decreasing resolution, etc.; they are caused not only by mismatched fittings, but also by improper fitting procedure or by using long tubing. When connecting the components from the IC system, it is necessary to consider that the tubing length and diameter of plumbing between the injector, column and detector have to be kept at the minimum, in order to minimize their volumes (Meyer, 2013; Snyder et al., 2011). Higher volumes in these connections lead to increased tailing of peaks and an overall decrease in the quality of a separation.

2.1.6 The chromatographic column

The chromatographic column is the place where separation occurs; the separations' performances and, hence, the achievement of the proposed goal depends entirely on the right choice of the column and of its proper operation.

A chromatographic column is, in fact, a tube filled with very small particles of stationary phase (column packing). IC columns are made from PEEK or stainless steel, being inert for the analytes and the mobile phase and are able to withstand high operating pressures (Figure 33); common column lengths are in the range of 5–250 mm, with internal diameters between 0.2 and 4.6 mm. Analytical IC columns operates at low flow rates (~0.2–1.5 mL/min), being able to face around 1,000 injections (depending much on the type of samples and operating conditions). The connection of a chromatographic column with the system is accomplished by end-fittings (Figure 29). Because the column is packed with small particles, a frit[6] is necessary to hold them

---

6 The frit is a small piece of sintered metal having very narrow pores (0.5–2 μm).

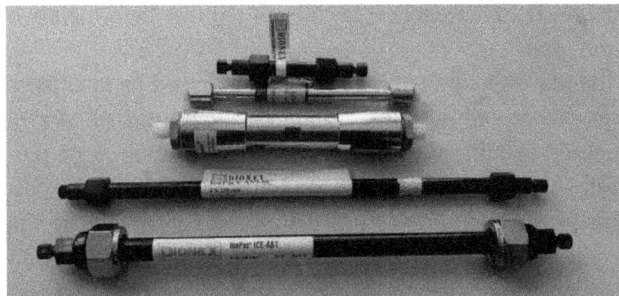

Figure 33: Typical IC columns.

inside. Each column has a flow arrow to indicate the direction in which it was packed; it has to be fitted in the system to provide the same flow through it as it indicates.

Selection of a proper column for a certain analytical approach can be a difficult task for a chromatographer, given the diverse range of columns available in the market. During recent years, the number of available columns has increased, with vendors offering different ranges of strong, weak or mixed-mode ion exchangers based on nonporous, porous and agglomerated resins, as well as silica-based phases. For selecting a chromatographic column for a given separation, it is important to consider several technical characteristics such as:

- the nature of the stationary phase, each column being designed for a certain type of separations;
- the column's length – the bigger it is, the higher the overall separation times and, hence, the mobile-phase consumption and the cost per separation;
- the column's internal diameter – the smaller it is, the smaller the mobile-phase consumption and the cost per separation;
- the particle sizes for the stationary phases (2–5 μm) – the smaller they are, the higher the efficiency.

A good starting point is to perform a literature search for a similar separation, and then, to adapt it as necessary. The cost factor is a major one in the buying decision; hence, it is necessary to consider that as the internal diameter and the length of a column increases, so does the consumption of the mobile phase that is necessary for a chromatographic analysis, as well as the duration of the separation; hence, the cost of the analysis.

On the other hand, the smaller the size of the particles from the column packing, the higher the efficiency, the higher the backpressure and also the column's price, but smaller diameters for the stationary-phase particle, shorter columns and small-diameter columns reduce analysis times and solvent consumption, while assuring higher sensitivity and lower limits of detection. When analyzing samples with a relatively simple composition, using shorter columns, while maintaining the same stationary-phase diameter is advantageous, because it reduces both the analysis time and solvent consumption.

Capillary columns are the narrowest (0.2–0.6 mm), being followed by microbore (1–3 mm) and standard bore columns (3–4.6 mm). Column packing consists of very small particles that are squeezed together very tightly inside the column; these particles are usually spherical in shape, having a porous structure. The size of the particles from column packing is very important, and it is always specified by the producer. Most columns come with particles of 5 μm, although both smaller and larger particles are used. The particle diameter of the column packing has an important effect on chromatographic separations and on pressure: the smaller the particles, the higher the efficiency (higher number of theoretical plates, higher resolution), but, on the other hand, a higher pressure is needed to maintain a given flow rate.

Since the columns are the main factors responsible for the quality of separations, they must be kept in top shape and should be treated with utmost care and to follow several recommendations:

- It is important to read the instructions from the column supplier and/or on-line information and to follow them, in order to extend as much as possible the column lifetime, since the column chemistry and the recommended mobile phases dictates the range of separations.
- Before a new column is first used, it should be thoroughly equilibrated with the mobile phase that is to be used.
- Column lifetimes should be higher than 1,000 injections if using clean samples, but to achieve this, traces of proteins and hydrophobic contaminants should be removed, samples have to be carefully filtered through membrane filters and using a guard column is recommended.
- On completion of the analysis, it is necessary to wash the columns with an appropriate mobile phase, followed by storage in recommended receipt, which will ensure the optimal storage conditions, while preventing microbial and algal growth.
- When not in use, the columns have to be kept air-tight (with column plugs in both ends to prevent the stationary phase drying out) and free from vibrations.
- It is useful to keep a record of each column in a laboratory journal, including test chromatograms, number of injections, types of samples, changes in the composition of the mobile phase, flow rate, temperature, user name(s), back-pressure, events; such a practice can help in diagnosis, when the quality of the separations achieved on a column decreases.

Being the central part of the chromatographic system, its proper operation is essential; fortunately, there are several devices that can assist a chromatographic column, protecting it from several destructive factors: guard columns, in-line filters, precolumns and trap columns, their positions relative to the chromatographic column being depicted in Figure 34 (Dolan, 2014; Nugent and Dolan, 2019).

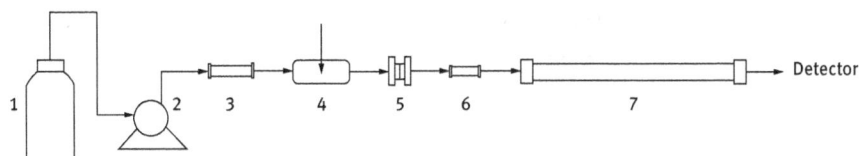

Figure 34: Possible protections for an analytical column: precolumn (3), in-line filter (5) and guard column (6). Other IDs: 1, solvent reservoir; 2, pumping system; 4, detector; 7, analytical column.

Guard columns are short columns (typically 1–5 cm in length) packed with the same stationary phase as the analytical columns, positioned between the sample injector (or an in-line filter) and the inlet of the analytical column, preventing undesirable components in the sample from entering the separation column and contributing to prolonging the lifetime of the analytical column (Figure 35). A guard column can also help avoiding a common problem: the contamination of the top of the analytical column by strongly retained compounds from the samples. The guard columns have to be discarded and replaced at intervals depending on the contamination level present in the samples being analyzed; for appropriate protection, they have to be replaced when 50% is exhausted (Dong, 2006).

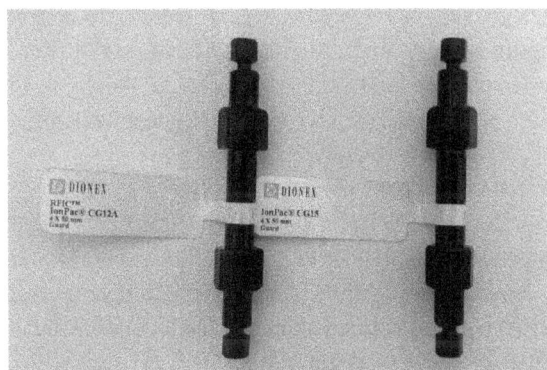

Figure 35: Guard columns.

Additionally, it is possible to use **trap columns** before the guard column when working with difficult samples; these are designed to retain certain components from the samples, enhancing the separation of those of interest (e.g., Dionex IonPac NG1 – trap for organics and humics, Dionex IonPac CP1 cation polisher – for very high concentrations of calcium and / or magnesium).

For a more effective protection of the analytical column, in-line filters can be used in conjunction with guard columns. **In-line filters** are cheap devices used to trap the particles that could plug the analytical column; they consist of a frit-holder containing a 0.5–2 µm frit, located between the sample injector and the inlet of the analytical column or guard column. When the frit becomes plugged by particles, an increase in operating pressure or a deterioration in the separation is observed, and the frit has to be replaced: for this, the frit-holder is opened, the plugged frit is replaced with a new frit and then, the frit-holder is reassembled.

In certain cases, **precolumns** are used to precondition the mobile phase, in order to minimize the effects of certain compounds from the mobile phase on the analytical column packing; since a precolumn is located between the pump and the sample injector, the sample does not pass through it.

2.1.7 The column oven

Column temperature is an important factor for both selectivity, efficiency and detection, and also has implications on possible misidentification of peaks. An increase in temperature generally leads to a decrease in the retention times of all peaks, and it is often accompanied by certain decrease in resolution or even in change of the relative positions of the peaks in the chromatogram. In a laboratory in which temperature varies, so will the retention times in chromatographic separations. In IC, the retention times are not much influenced by temperature, but an accurate temperature control is a must when using conductivity detectors or refractive index detectors, since temperature variations can cause important drifts and compromise the quantification. Hence, for obtaining reproducible results, the analytical column has to be placed in an efficient thermostating system in order to maintain a constant temperature during separation (Colin et al., 1986; Dong, 2006; Weiss, 2016).

Several technical solutions are commercially available; they differ in the way they transfer heat to the column, whether they heat only, or they heat and chill (some systems can achieve subambient temperature, using Peltier devices), the accuracy and precision with which they control temperature, their capacity (most can host several columns, some of them being able to perform column switching using special valve units), the most common being block heathers and column ovens.

Block heathers transfer heat from a heating element located in a metal block on which the column(s) fit tightly; such a design provides +/−1 °C accuracy, at best; they are more popular, yet, these have been reported to impart uneven heating. **Column ovens** use forced air circulation by a fan-assisted stream and can heat up and cool down more rapidly, assuring an accuracy of up to +/−0.1 °C in maintaining a constant temperature.

2.1.8 The suppressor

Suppression is an important stage in the analysis of ionic species by IC using conductivity detection, because it enables the reduction of the conductivity of the mobile phases. Such an approach is necessary when dealing with large concentration of ions in the mobile phase, which makes it difficult to detect low concentrations of certain ionic solutes. There are two alternatives to reduce the background conductivity: electronic suppression and chemical suppression (Fritz and Gjerde, 2010; Schaefer et al., 1989; Weiss, 2016).

Electronic suppression is a common method consisting of the baseline change, accomplished usually by the instrument's software that acts similar to the auto-zero function of spectrophotometers. When this function is activated, the baseline becomes stable at a conventional zero conductivity level. IC with electronic background suppression is often referred to as **direct IC** or **nonsuppressed IC**, since the eluent leaves the separating column, without any alteration in composition; this technique is effective when using mobile phases with a relatively low background conductivity.

Chemical suppression is accomplished using a specialized device (a suppressor) placed after the separation column, which reduces the background conductivity of the eluent, while enhancing the conductivity of the analytes. In this case, suppression acts by ion exchange, both the mobile phase composition and the analyte(s) being affected: certain ions from the eluent are changed with others, with lower conductivity, thus decreasing the background conductivity of the eluent and minimizing the baseline noise, and hence, increasing the signal-to-noise ratio. As a result, the background noise is reduced, increasing the sensitivity and reproducibility (Figure 36). Chemical suppression is based on the use of salts of weakly dissociating acids as eluents, which are later eliminated to a large extent by a second ion-exchange step.

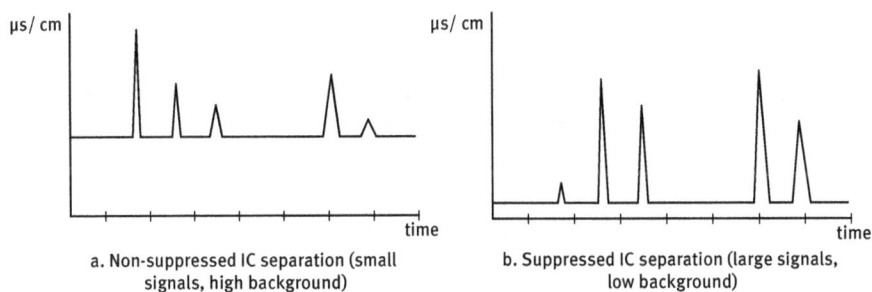

a. Non-suppressed IC separation (small signals, high background)

b. Suppressed IC separation (large signals, low background)

Figure 36: Nonsuppressed (a) versus suppressed (b) IC separations.

In anion analysis, the suppressor is a cation exchanger, which replaces the cations from the eluent with protons, transforming the analytes in free anions with protons as counter ions, thus resulting in a high increase of the conductivity signal. If the case of chloride is considered, this anion can have sodium as a counter ion and this, during suppression, is replaced with a proton with a much higher conductivity, thus, increasing the signal strength:

$$\text{Na}^+ + \text{Cl}^- + \text{RSO}_3 - \text{H}^+ \rightleftharpoons \text{H}^+ + \text{Cl}^- + \text{RSO}_3 - \text{Na}^+$$

low conductivity   high conductivity

When using sodium carbonate/sodium hydrogen carbonate as eluent, the suppressor also exchanges sodium cations from it with protons forming carbonic acid, which is unstable and only weakly dissociated, decreasing the background conductivity:

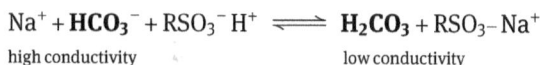

$$\text{Na}^+ + \text{HCO}_3^- + \text{RSO}_3^- \text{H}^+ \rightleftharpoons \text{H}_2\text{CO}_3 + \text{RSO}_3 - \text{Na}^+$$

high conductivity   low conductivity

The conductivity of the eluent can be reduced even more by also suppressing carbon dioxide by means of a CO_2 suppressor using membrane technology, which acts by decomposing the carbonic acid and removing the formed carbon dioxide under vacuum, hence leaving only water, with a very low conductivity:

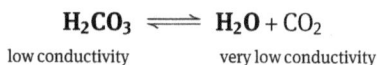

$$\text{H}_2\text{CO}_3 \rightleftharpoons \text{H}_2\text{O} + \text{CO}_2$$

low conductivity   very low conductivity

In cation analysis, the suppressor is an anion exchanger, which replaces all anions with hydroxide ions. Considering the case of calcium cation, this can have chloride as a counter ion; during suppression, chloride is replaced with HO^- with a much higher conductivity, thus increasing the signal strength:

$$\text{Ca}^{++} + \text{Cl}^- + \text{R} + \text{OH}^- \rightleftharpoons \text{Ca}^{++} + \text{HO}^- + \text{R} + \text{Cl}^-$$

lower conductivity   higher conductivity

When using azotic acid as eluent, the suppressor also exchanges anions with HO^-, which reacts further with protons generating water, with very low conductivity, decreasing the background conductivity:

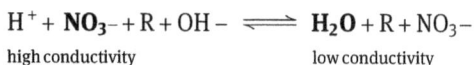

$$\text{H}^+ + \text{NO}_3- + \text{R} + \text{OH} - \rightleftharpoons \text{H}_2\text{O} + \text{R} + \text{NO}_3-$$

high conductivity   low conductivity

Chemical suppression is more advantageous for anion determination, significantly increasing the sensitivity than in cation determinations. Unfortunately, the choice of the mobile phases when working with chemical suppression is limited (usually, alkali hydroxides and mixtures of alkali carbonates and hydrogen carbonates).

Chemical suppression can be accomplished using either column-based or membrane-based suppressors. Suppressor columns need to be periodically regenerated, but automatic systems using suppressor columns can switch between these, so that one can be regenerated while another one is in use. Membrane-based suppressors are more advantageous, since they allow continuous regeneration.

Nonsuppressed IC is easier to perform and cheaper; food analysis can be performed in nonsuppressed mode, because the quantitation levels are usually in the upper mg/L up to low percentage levels. However, suppressed IC can prove to be useful in research, since the use of suppressors gives a lower background and hence, a higher signal to noise ratio and a better sensitivity.

2.1.9 The detector

The detector is a device continuously measuring a certain property of the eluent that leaves the chromatographic column and passes through it, producing an electric signal proportional to the concentration of each separated analyte. When no analyte is passing through the detector, the measured output is a constant signal – a **baseline**. When an analyte reaches the detector, the detector responds, giving rise to a change in the output signal, seen as a **peak**.

Detectors can be broadly classified into universal and selective detectors (Swartz, 2010):

- **Universal detectors** continuously monitor a bulk property of the mobile phase, which changes as an analyte elutes (e.g., conductivity and refractive index); they have a certain signal output in the absence of a solute leading to serious limitations, since the addition of a small amount of a solute will cause an increase to the background signal, so that these detectors generally have poor limits of detection. Since the universal detectors also respond to the mobile phase, their response changes with variations in the mobile-phase composition and these detectors are incompatible with gradient elution techniques.
- **Selective detectors** respond to a certain property of the analytes (e.g., ultraviolet-visible absorbance detectors, electrochemical detectors) and generally have much lower limits of detection, but are applicable only for monitoring several compounds exhibiting that specific property.

A detector has several important characteristics to consider when deciding a configuration for an IC system (Foley and Dorsey, 1984), from which:

- **Sensitivity** – the response per unit concentration, being usually determined from the slope of the calibration curve and reported in units of signal/amount of an analyte; it depends on the noise level (in general, the more universal a detection method is, the higher its noise level and the lower its sensitivity).
- **Selectivity** – the ability of a detection method to be sensitive to certain species and insensitive to others.
- **Linearity** – the ability of a detector to produce a proportional response with the amount of the analyte over the working range of concentrations.
- **Linear range** – the range of concentrations in which the response of a detector is proportional to the concentrations injected.

- **Drift** – a deviation of the baseline with time due internal factors, which appears in data acquisition as a slope; the drift is normal when an IC system is switched on, being necessary to allow at least 30 minutes the system to stabilize.
- **Noise** – a random change of the baseline signal, which occurs regardless of any input signal, in the presence of mobile-phase flow, caused by a wide range of contributors (electronics, the power supply, impurities in the mobile phase, bubbles, changes in temperature, etc.); it should be as low as possible, since it determines the limits of detection.
- **Signal-to-noise ratio (S/N)** – the ratio between the wanted signal and the unwanted background noise.
- **Limit of detection (detection limit, LOD)** – the amount of analyte that can be distinguished with some level of certainty from the baseline noise (Vial and Jardy, 1999); usually, it is defined as that amount of analyte giving a signal three times the standard deviation of a blank (three times the noise).
- **Response time** – the time a detector takes to respond to a change in the analyte's concentration in the detector cell. It can modify both the peak shapes and peak heights, causing an apparent change in both column efficiency and detector sensitivity. A proper adjustment of the response time can have a dampening effect (detectors usually have a user-variable time constant), reducing the noise; the bigger its value, the more important the dampening effect, but peaks can become wide and tailed. If it is too small, the noise level is higher and the limit of detection may be adversely affected.
- **The flow cell volume** is an important constructive issue for both maximum detector sensitivity and to prevent loss of peak resolution obtained from the column; the flow cell volumes are in the range of 0.02–20 μL, depending on the detector type, while related to the column's type (smaller flow cells for systems equipped with capillary columns, larger for standard bore columns).

There is no detector that can fulfill all the requirements for a particular determination, each of them having strengths and weaknesses; the choice of a certain detection technique in an IC system is determined mainly by the characteristics of the separation technique and the properties of target analytes.

2.1.9.1 Conductivity detectors

Conductivity detectors are the most commonly used in IC, conductivity detection being one of the oldest of the LC techniques and an universal detection method for ionic analytes (Weiss, 2016).

The electrical conductivity of a solution is the measure of its ability to carry an electrical current. The conductivity of a sample solution is an additive property, depending on the type of ions, concentrations of dissolved ions and their electrochemical properties (Skoog et al., 2013):

$$\kappa = \sum_i c_i \cdot Z_i \cdot \lambda_i$$

where:

c – concentration (mol/L);

Z – the charge of ionic species;

λ – the equivalent conductivity (S cm^2/mol).

The conductivity is a nonspecific universal sum parameter caused by all dissolved ionic species (salts, acids, bases and some organic substances) from a solution; since it is measured as a bulk property for a given system, it is unable to differentiate between diverse ions, the obtained value being proportional to the combined effect of all ions from the sample.

Conductivity can be measured as a reciprocal of the resistance of a solution placed between two electrodes:

$$\kappa = \frac{L}{A \cdot R}$$

where:

L – the distance between electrodes;

A – the area of the electrodes;

R – the electrical resistance of the solution.

The equivalent conductivity is a specific attribute of every ionic species (Table 3), depending not only on ion type, but also on concentration and temperature. Temperature influences the equivalent conductivity in a different manner: a higher temperature increases the motion of the ions and lowers the water viscosity, leading to increased mobility of the ions and a higher conductivity. In the case of weak electrolytes, a higher temperature raises the dissociation rate and therefore, also the conductivity. As errors caused by changes in temperature are extremely visible, temperature fluctuations within the conductivity detector should be no larger than +/−0.01 °C.

As shown in Table 3, certain ions are good charge carriers (e.g., H$^+$ or HO$^-$, with small sizes and high mobility), while others exhibit lower conductivity (e.g., magnesium or phosphate). The higher the concentration, the larger the number of charge carriers and hence, higher the conductivity.

Conductivity detectors are devices designed to measure the difference in background conductivity of the eluent versus the conductivity of the ion species separated by the analytical column, using a flow cell with two electrodes, across which an alternating current potential is applied (Figure 37). Since the sensitivity of these detectors depends on the temperature, the temperature must be kept strictly constant (+/−0.01 °C) in the conductivity flow cell. Their sensitivity is high and they are

Table 3: Equivalent conductivities for common inorganic ions
(Foley and Haddad, 1986; Kott, 2019).

Cations	Λ_∞ (Scm²/mol)	Anions	Λ_∞ (Scm²/mol)
H^+	349.8	OH^-	198.6
Li^+	38.7	F^-	54
Na^+	50.1	Cl^-	76.4
K^+	73.5	HCO_3^-	44.5
NH_4^+	73.5	NO_2^-	72
$\frac{1}{2} Mg^{2+}$	53.1	NO_3^-	71.5
$\frac{1}{2} Ca^{2+}$	59.5	$\frac{1}{2} SO_4^{2-}$	80.0
$\frac{1}{2} Ba^{2+}$	64	$1/3\ PO_4^{3-}$	69

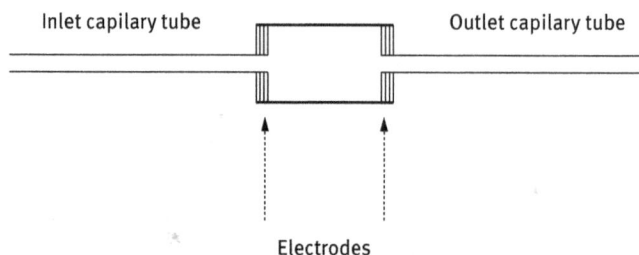

Inlet capilary tube   Outlet capilary tube

Electrodes

Figure 37: Schematic diagram of a conductivity detector flow cell.

universal for ionic species; they are not compatible with gradient elution. Usually small changes in conductivity occur when an analyte is eluted against an important background signal; however, because the common noise is small, sensitivity is very high (Foley and Haddad, 1986; Morgan and Smith, 2010).

Anions are usually analyzed using a low conductivity eluent, prepared using salts of phthalic, benzoic or salicylic acids; when an anion with a higher ionic conductivity appears in the detector's flow cell, the conductivity increases, and a peak is obtained. For alkali and alkaline earth cations, separations are usually accomplished using dilute acid eluents (2–3 mM), having very high conductivity (due to the very high equivalent ionic conductivity of protons); hence, when a cation with a lower ionic conductivity reaches the detector's flow cell, the conductivity decreases and negative peaks are obtained. For cation analysis all peaks are, therefore, negative, making it necessary to invert the detector signal to get positive peaks (Gjerde et al., 1979; Scott, 1996; Swartz, 2010).

The major advantages of conductivity detectors (high sensitivity, simplicity, no maintenance required) make them the preferred option in many applications in food analysis, both in nonsuppressed systems (Table 4) and in suppressed ones

Table 4: Selected IC applications in food analysis using nonsuppressed conductivity detection.

Analytes/matrix	Column	Mobile phase	Separation time (min)	Reference
Anions (dihidrophosphate, chloride, bromide, nitrate, sulfate) and organic acids (acetic, succinic, pyroglutamic, lactic, pyruvic, oxalic, citric) in beer	Shodex IC-I524 A	1.5 mM phthalic acid + 1.38 mM tris (hydroxymethyl) aminomethane + 300 mM boric acid 1.2 mL/min	20	Shodex application note – anions in beer
Anions (chloride, nitrate, sulfate and phosphate) in beer	Metrohm A SUPP 5	$Na_2CO_3/NaHCO_3$	24	Bruce (2002)
Cations (sodium, potassium, magnesium, calcium) in beer	Metrosep C 6	2.3 mM HNO_3 + 1.7 mM dipicolinic acid 0.9 mL/min	20	Metrohm, IC application note D-002
Cations (sodium, ammonium, potassium, magnesium, calcium) and monoethanolamine in red wine	Shodex IC YS-50	4 mM methanesulfonic acid 1 mL/min	10	Shodex application note YS-50
Anions (hydrogen carbonate, chloride, nitrate, hydrogen phosphate) in milk	Shodex IC I-524A	1.5 mM p-hydoxybenzoic acid + 1.7 mM N,N-diethylethanolamine + 10% CH_3OH 1.5 mL/min	20	Shodex application note – anions in milk

(Table 5), the latter being more sensitive. The lack of selectivity of the conductivity detectors is their greater limitation; because conductivity is a cumulative parameter of the ions' concentration from a sample, it does not allow discrimination between the type of ions, the differences in the equivalent conductivities of those ions being too small to be analytically exploitable.

Table 5: Selected IC applications in food analysis using suppressed conductivity detection.

Analytes/matrix	Column	Mobile phase	Separation time (min)	Reference
Fluoride, chloride, nitrate, sulfate in water	Metrosep A Supp 5–250/4	3.2 mM Na_2CO_3/1 mM $NaHCO_3$ 0.7 mL/min	14	Metrohm IC application note S-287

Table 5 (continued)

Analytes/matrix	Column	Mobile phase	Separation time (min)	Reference
Fluoride, chloride, nitrite, nitrate, selenate, arsenate, perchlorate, chromate in water	Metrosep A Supp 7	10.8 mM Na_2CO_3/35% acetonitrile 0.8 mL/min	20	Mohana Rangan et al. (2021)
Fluoride, chlorite, bromate, chloride, nitrite, bromide, chlorate, nitrate, phosphate, sulfate in water	Shodex IC SI-52 4E, 250/4	5.4 mM Na_2CO_3 0.8 mL/min	20	Shimadzu application news HPLC-022
Fluoride, chloride, bromide, nitrate, phosphate, sulfate, chromate in water	Dionex IonPac AS1	5 mM Na_2CO_3 in 1 mM $NaHCO_3$ 1 mL/min	14	Konczyk et al. (2018)
Fluoride, chloride, nitrite, nitrate, phosphate, sulfate in water	Dionex IonPac AS14A (3 × 150)	8 mM Na_2CO_3/1 mM $NaHCO_3$, 0.8 mL/min	6	Thermo Fisher App. note 140
Anions and organic acids (fluoride. lactate, acetate, pyruvate, chloride, nitrate, succinate, malate, sulfate, phosphate, citrate) in ale	IonPac AS11	Water/1 mM NaOH/ 100 mM NaOH/methanol 2 mL/min	20	Thermo Scientific application note 46
Cations (potassium, sodium, magnesium, calcium) in lager	IonPac CS12	Water/100 mM methanesulfonic acid 1 mL/min	12	Thermo Scientific application note 46
Oxalic acid (+ fluoride, chloride, nitrate, sulfate and phosphate) in beer	Allsep Anion 100	0.85 mM $NaHCO_3$, 0.9 mM Na_2CO_3 1.2 mL/min	17	Marten (2003)

Table 5 (continued)

Analytes/matrix	Column	Mobile phase	Separation time (min)	Reference
Anions (chloride, phosphate, nitrate and sulfate + sulfite) in beer	Metrosep A Supp 10	4 mM Na_2CO_3 + 6 mM $NaHCO_3$ + 5 µM $NaClO_4$ 0.7 mL/min	17	Metrohm IC application note D-002
Sulfite (+chloride, phosphate, sulfite, bromide, nitrate and sulfate) in beer	Metrosep A Supp 10	6 mM Na_2CO_3 4 mM $NaHCO_3$ 5 µM $NaClO_4$ 0.7 mL/min	22	Metrohm IC application note S-225
Anions (fluoride, chloride, nitrate, phosphate, sulfate) in beer	Dionex Ion Pac AS 14 (250 × 4.6 mm)	3.5 mM Na_2CO_3 + 1 mM $NaHCO_3$ 1 mL/min	–	Michalski et al. (2021)
Organic acids and anions in wine	Metrosep A Supp 16	1 . . . 60 mM NaOH (gradient) 0.7 mL/min	100	Metrohm IC Application Note S–362.
Organic acids (citric, tartaric, malic, succinic, lactic, fumaric, acetic) + carbonate in wine	2x HPICE-AS1 in series	2 mM octanesulfonic acid in 2% 2-propanol 0.5 mL/min	60	Thermo Scientific/ Dionex – application note 21
Organic acids (citric, tartaric, malic, succinic and lactic, acetic) + carbonate in wine – fast run	HPICE-AS1	2 mM octanesulfonic acid in 2% 2-propanol 0.8 mL/min	15	Thermo Scientific/ Dionex – application note 21
Acetate, chloride, phosphate, malate, **sulfite**, tartrate, sulfate and oxalate in wine (fast screening)	Metrosep A Supp 10	5 mM Na_2CO_3 + 5 mM $NaHCO_3$ + 5 µM $HClO_4$ 1 mL/min	18	Metrohm IC Application Notes S–281, S-396
Lactate, chloride, nitrate, **sulfite** and phosphate in wine	Metrosep Anion Dual 2	2 mM $NaHCO_3$, 1.8 mM Na_2CO_3, 15% acetone 0.8 mL/min	18	Metrohm IC Application Note S–12
Lithium (+sodium, potassium, magnesium and calcium) in wines	IonPac CS12	20 mM methanesulfonic acid 1 mL/min	15	Zerbinati et al. (2000)

Table 5 (continued)

Analytes/matrix	Column	Mobile phase	Separation time (min)	Reference
Sulfite + anions (chloride, nitrate, hydrogen phosphate, sulfate) in wine	Shodex IC SI-90 4E	1 mM Na_2CO_3 + 4 mM $NaHCO_3$ + 5% Acetone 1.5 mL/min	18	Shodex application note SI-90 4E
Iodide in milk powder	Metrosep A Supp 5	1 mM $NaHCO_3$ 3.2 mM Na_2CO_3 0.7 mL/min	30	Metrohm IC application note S–162
Choline in dry milk and infant formula	IonPac CG12A	18 mM H_2SO_4 1 mL/min	25	Dionex/ Thermo Scientific, Application Note 124
Choline (+sodium, ammonium, potassium, calcium, magnesium) in baby milk powder	Metrosep C Supp 1	4 mM HNO_3, 50 µg/L rubidium 1 mL/min	28	Metrohm IC application note CS-004
Iodide, thiocyanate and perchlorate in milk	Metrosep A Supp 15–50/4.0	4 mM Na_2CO_3 /6 mM $NaHCO_3$, 10% MeOH 0.8 mL/min	28	Metrohm IC Application Note S–297
Anions (chloride, sulfate, bromide, phosphate, citrate and isocitrate) in milk whey	IonPac AS11-HC	20 mM NaOH 0.38 mL/min	15	Cataldi et al. (2003)
Cations (lithium, sodium, ammonium, potassium, calcium, magnesium) in milk whey	IonPac CS12A	20 mM methanesulfonic acid 1 mL/min	10	Cataldi et al. (2003)
Ammonium (+ lithium, sodium, ammonium, potassium, calcium, magnesium) in milk and dairy products	IonPac CS15	5 mM H_2SO_4 + 9% acetonitrile 1.2 mL/min	40	Gaucheron and Le Graet (2000)

2.1.9.2 Ultraviolet-visible absorbance detectors

Ultraviolet-visible (UV-VIS) absorbance detectors measure the interactions of ultra-violet (190–400 nm) or visible light (400–780 nm) with the separated ions. When light radiation passes through a sample solution, a part of it is absorbed, and by using a spectrophotometer it is possible to measure this interaction (Figure 38).

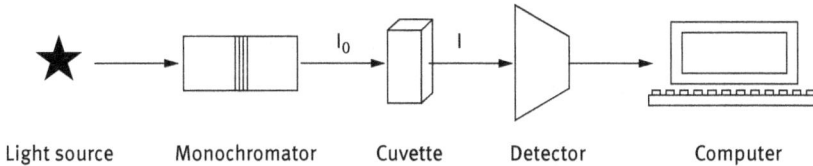

Light source   Monochromator   Cuvette   Detector   Computer

Figure 38: A simplified schematic representation of a spectrophotometer.

Light absorption depends on the nature of the analyte, the wavelength and the concentration and can be expressed in units of transmittance or absorbance:

$$T = \frac{I}{I_0}$$

$$A = -\log T$$

where:
T – transmittance;
A – absorbance;
I_0 – the amount of light entering the sample solution;
I – the amount of light living the sample solution.

According to the Berer's law, the absorbance of a solution is directly proportional to its concentration:

$$A = e.b.c$$

where:
A – the absorbance;
e – the molar absorptivity;
b – the pathway length of light in the solution;
c – the concentration of absorbing molecules.

Absorptivity changes with wavelength and hence, a condition for the application of Beer's law is to use monochromatic light in measurements.

Since absorbance is unitless, formal absorbance units (AU) are used for expressing it. The y-axis in the chromatograms generated using UV-Vis detectors is labeled in AU, often expressed as "Absorbance Units Full Scale" (AUFS).

UV-VIS detectors were among the first detectors used for liquid chromatography, but in IC, they can be used only in cases where the analytes absorb electromagnetic radiations in the UV-VIS range (e.g., aminoacids, organic acids, bromide, chromate, iodate, iodide, nitrate, nitrite, thiosulfate and certain metal complexes). UV-VIS detection can be used in IC systems as direct detection technique, indirect detection technique or in conjunction with post-column derivatization (Buchberger, 2001; Swartz, 2010).

Direct UV-VIS detection can be used for the determination of species that absorb in the UV range (Table 6); this approach is helpful when the target ions are in a matrix containing very high amounts of other ions (e.g., chloride, phosphate, and sulfate), which have no UV absorption (Marshall et al., 2017). In this way, the differences in the absorption behavior of certain analytes can be used to avoid interferences with analytes of interest (e.g., in samples having high salt content, very low amounts of nitrite and nitrate can be determined).

Table 6: The maximum absorption wavelengths of certain chromophores (Birkman et al., 2018).

Chromophore	Λ_{max} (nm)
Br^-	225
BrO_3^-	195
I^-	195
NO_2^-	214
NO_3^-	203
R-COOH	200
R-CHO	210

Indirect UV-VIS detection can be applied in IC, when using mobile phases with high UV absorption (e.g., phthalate-Λ_{max} 300 nm, benzoate-Λ_{max} 254 nm), so that the analytes reduce the detector signal; in such cases, ions with low or no UV absorption will generate negative peaks, while ions with higher UV absorption will cause positive peaks (Buchberger, 2001).

UV-VIS detection with post-column derivatization can be used for the detection of transitional metals (Ding et al., 2000; Metrohm IC Application Note C-105; Shodex application note KC 811; US EPA Method 218.7) for this approach, in the eluent stream, a post-column reagent is added and, then, after passing through a reaction chamber, the metal complexes are detected in the detector (Figure 39).

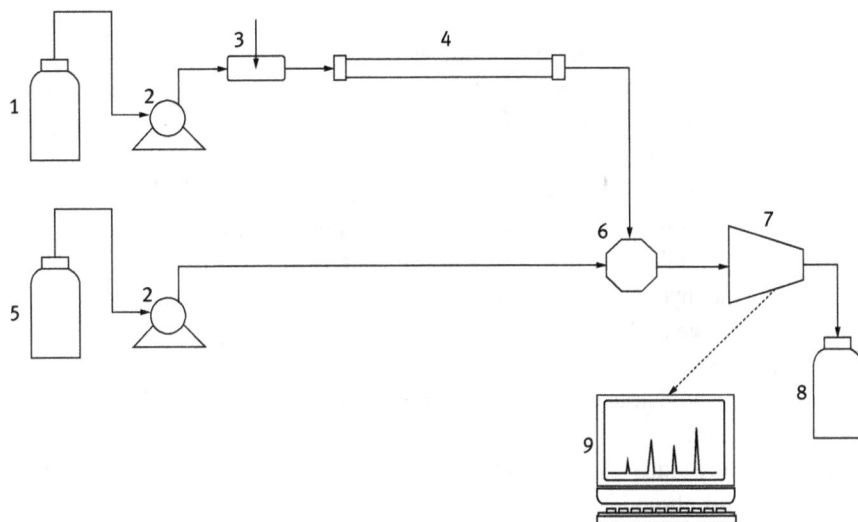

Figure 39: Schematic diagram of an IC system with post-column derivatization: 1, mobile phase reservoir; 2, pumps; 3, injector; 4, chromatographic column; 5,reagent reservoir; 6, post-column reactor; 7, detector; 8, waste reservoir; 9, computer.

A growing number of applications using post-column derivatization[7] in combination with spectrophotometric detection opened the field of transition metal analysis for IC, thus providing a powerful alternative to conventional atomic spectroscopy methods. Post-column derivatization will not change the separation, but involves the use of a supplementary pump to introduce the derivatization reagent to the effluent from the column via a derivatization module, where a chemical reaction occurs. In such cases, chelation of the metal ions eluted from the chromatographic column with a color-forming reagent occurs before detection within visible wavelength range; a common reagent is 4-(2-pyridylazo)-resorcinol (PAR), which can react with a large number of metal cations, forming chelates that are later detected at 510–530 nm (Klingenberg and Seubert, 1998). Post-column derivatization can be successively applied also in the case of anion's determination (Michalski, 2003; Michalski and Lyko, 2013; Michalski and Lyko, 2010); hence, bromate can be determined by using the triiodide method, based on the following reactions that occur in an environment containing sulfuric acid:

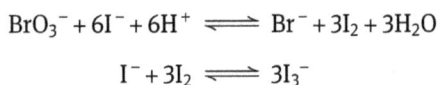

$$BrO_3^- + 6I^- + 6H^+ \rightleftharpoons Br^- + 3I_2 + 3H_2O$$

$$I^- + 3I_2 \rightleftharpoons 3I_3^-$$

---

[7] *Derivatization* involves the conversion of an analyte into a different compound having more favorable properties to be determined in a given analytical context. In IC, an analyte with no chromophore (hence with no possibility to be detected by UV-VIS detection) can be converted in an adsorptive species.

The reaction requires molybdate ions as catalyst, the resulted triiodide being finally detected at 325 nm. Determination of aminoacids is another area of applications in which post-column derivatization is highly appreciated (Haddad and Jackson, 1990); although aminoacids are easily resolved using IEC, the common detection methods are not efficient for them (most aminoacids have little or no UV absorption and cannot be oxidized amperometrically, while conductivity detection is not sensitive enough). By derivatization with ninhydrin, these compounds can be detected using VIS detection, but a relatively high temperature is necessary; instead, derivatization with o-phthalic aldehyde can be accomplished at room temperature (Kriukova et al., 2019). A disadvantage of this technique lies in the increased complexity of the IC system, which also involves a cost factor.

The common UV-visible detectors are variable-wavelength detectors and photo-diode-array detectors.

Variable-wavelength UV-VIS detectors consist of a light source, a monochroma-tor, a flow cell and photodetector (Figure 40). The light radiation originating from a lamp is focused on the entrance slit of a monochromator and, then, a diffraction grating separates the different wavelengths; another slit follows to select the desired wavelength, after which the monochromatic light is directed to the flow cell and, finally, onto a photomultiplier tube or a photodiode, where its intensity is converted into an electrical signal. UV–VIS detectors typically cover a wavelength range from UV to near infrared (190–800 nm), requiring two radiation sources: a deuterium lamp for UV range and a tungsten lamp for visible spectral range. The flow cell is the place where the light adsorption occurs; the transparent components of this device are made of quartz, to allow both low and high wavelength radiations to pass through these. When there is no UV-VIS absorbing analyte in the detector flow cell, light passes through it and generates a maximum electrical signal; when an absorbing analyte is present in the flow cell, it reduces the amount of light reaching the sensor and causes a change in the electric signal, which is electronically inverted, resulting a positive peak in the chromatogram (Dong and Wysocki, 2019; Kraiczek et al., 2014).

Variable-wavelength detectors can improve sensitivity by operating at the wavelength of maximum absorbance of the analytes; when required, they can operate at wavelengths where other solutes will not absorb, thus avoiding possible interferences. The wavelength's choice is crucial: the working wavelength should be set at maximum absorbance or the analytes (Λ_{max}) or close to it, to have the strongest possible absorption; using an inappropriate wavelength may result in small peaks or even no peaks at all. Since a single wavelength can be selected, all the analytes will be detected at this one, and hence, a compromise is a must (not all the analytes have the same Λ_{max}, since they have different absorbance spectra – a small amount of an analyte that absorbs strongly at the selected wavelength can give a bigger peak than a large amount of a weak absorber). If an interfering peak is present, it can be excluded selectively by choosing a wavelength at which it does not absorb, as long as other peaks of interest still have some absorption (Swartz, 2010).

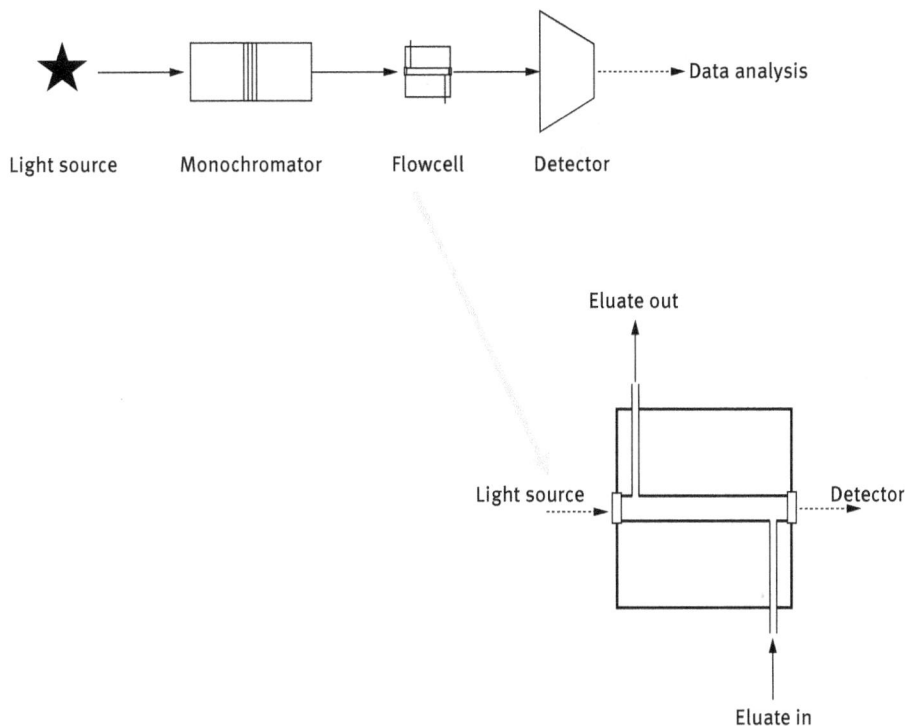

Light source   Monochromator   Flowcell   Detector

Eluate out

Light source   Detector

Eluate in

Figure 40: Schematic diagram of a variable wavelengths UV-VIS detector (a), with a detail of the flow cell.

Diode array detectors (DAD) monitor the entire spectrum versus time, instead of just one wavelength as with common UV-VIS detectors; they are one among multidimensional detectors, since they produce three-dimensional chromatograms with time, wavelength and absorption axes. In such a device, polychromatic radiation passes through the flow cell, being dispersed by a grating and falls into an array of photodiodes, each measuring a narrow band of wavelengths of the spectrum at once (Figure 41). This is a so-called "reversed optics design" because the whole range of light wavelengths passes through the flow cell before being diffracted (Swartz, 2010).

DADs provide a fast parallel data acquisition, leading to three-dimensional data structures. By simultaneously collecting the entire spectra over a desired wavelength range, this type of detector offers multiple advantages:
- every compound can be quantified at its wavelength of maximum absorbance, hence providing maximum sensitivity for each component, which is important for trace analysis
- allow the molecular absorption spectrum be used as a supplementary criteria for peak identification, since the acquired spectra can be subsequently compared with reference compounds' spectra

- the spectra of all peaks are available as "time slices," regardless of how many peaks are present
- it is possible to establish if co-elutions occur, by using the contour plot (a three dimensional plot of absorbance, wavelength and time) or by comparing the spectrum of the leading edge versus trailing edge of a peak
- having no moving parts, this design is more robust and more precise than that of a classic UV-VIS detector.

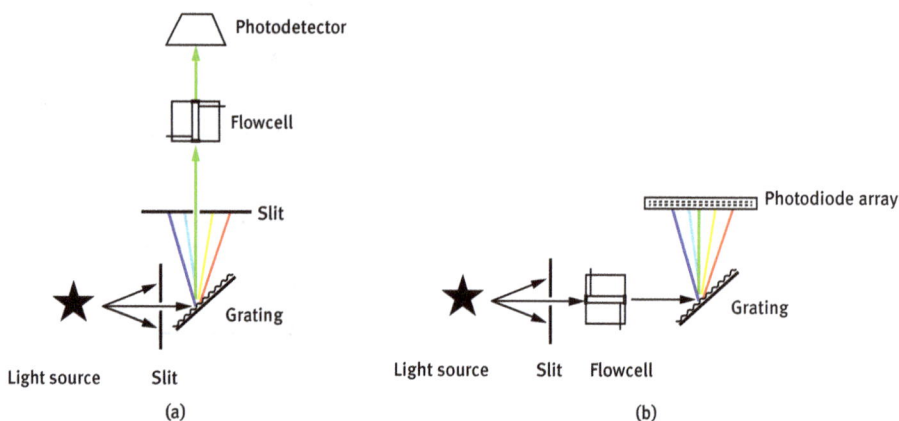

Figure 41: Simplified scheme of a variable wavelength UV-VIS detector (a) versus a photodiode array detector (b).

DADs have been increasing in popularity during the last few decades as a result of both improvements in their design and in the software for post-chromatographic data analysis; they are the best option when it is intended to simultaneously detect light absorption at different wavelengths.

UV-VIS detectors are highly sensitive, selective, nondestructive, have a wide linear range, are suitable for gradient elution, are not affected by changes in temperature and in flow rate and overall are easy to operate; these are important reasons for which they were selected in certain applications dealing with the analysis of wine or beer (Table 7).

2.1.9.3 Refractive index detectors

When light passes the boundary between two transparent layers, its velocity as well as its direction changes, causing the light beam to be bent (refracted); the light bending ability of a transparent medium is expressed by a dimensionless number – its refraction index. When a solute is dissolved in water, the refraction index is increased by an amount which is proportional with the concentration of the solute – this being the principle behind quantitative applications. When more solutes are dissolved, each of them

Table 7: Selected IC applications of UV-VIS detection in food analysis.

Analytes/matrix	Column	Mobile phase	Detection	Separation time (min)	Reference
Copper, zinc, iron (II) and manganese in wine	Metrosep C 2	1.75 mM oxalic acid + 2 mM ascorbic acid 1 mL/min	UV/VIS (after post-column reaction with 0.2 mM PAR, 2.4 M ammonia, 1 M acetic acid)	12	Metrohm IC Application Note C-105.
Organic acids (citric, pyruvic, gluconic, malic, succinic, lactic, fumaric, acetic, pyroglutamic) in beer	Shodex RSpak KC-G 6B + KC-811	4.8 mM $HClO_4$ 1 mL/min	VIS (430 nM, post-column derivatization)	30	Shodex application note KC 811

contributes to the refractive index of the solution and, hence, as in the case of conductivity measurements, it is not possible to use this approach for qualitative analysis.

Differential refraction index detectors (or refractive index detectors – RID) measure the difference between the refractive index of the mobile phase alone and that of the eluent, as it emerges from the column. Since all liquids have a certain refractive index and these change when a solute is dissolved in a liquid, a change in the refractive index indicates elution of analytes from the column (Kromidas, 2016).

In principle, a RID includes a light source, a flow cell and a photodiode (Figure 42).The flow cell has two compartments – a measurement one and a reference one; the system is equilibrated before performing separations, leading to identical content in the two compartments. During the separation, as analytes elute from the

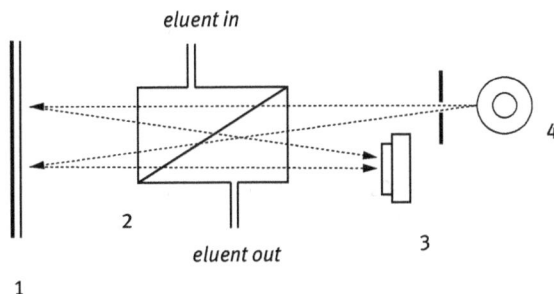

Figure 42: Schematic diagram of a refraction index detector: 1, mirror; 2, flow cell; 3, photodetector; 4, light-emitting diode.

column, changes in composition of the mobile phase occur, causing changes in the refraction index. It is important that the reference cell is flushed long enough for it to be filled with the mobile phase and be at constant temperature (the refractive index is temperature-dependent).

A major advantage of refractometric detectors is universality; they respond to changes in the refractive index of the mobile phase, though this changes not only with the solute concentration but also with temperature, pressure, dissolved gases and changes in mobile-phase conditions. Therefore, they are generally applicable only for isocratic separations; to be operated at maximum sensitivity, they need careful control of temperature and flow. When using a RID, it is important to allow enough time for temperature stabilization and to purge the reference cell properly (Al-Sanea and Gamal, 2022; Dolan, 2012).

Refractometric detectors are advantageous because of their ease of use: there are no wavelengths or voltage to set, while the maintenance is cheap (the lamp is inexpensive and lasts around 50,000 h).

Unfortunately RID have some drawbacks, such as:

– relatively small sensitivity, hence they are not appropriate for trace analysis;
– not usable with gradient elution, because even a small change in the composition of the mobile phase causes baseline drift (running a gradient causes important changes in the eluent refraction index, which can cause the baseline to drift off scale);
– high sensitivity to temperature fluctuations (any change in temperature causes baseline drift), a good temperature control being a must for both detector and column, so that a very good column oven with a very accurate temperature control is essential.

Table 8: Selected IC applications of refractive index detection in food analysis.

Analytes/matrix	Column	Mobile phase	Separation time (min)	Reference
Ethanol, maltotriose, glucose, fructose, lactic acid, glycerol in nonalcoholic beer	Aminex HPX 42A	Water 0.5 mL/min	21	Klein and Leubolt (1993)
Lactose, glucose, fructose, lactic acid in lactic acid beverages and fermented milk	Shodex sugar SH-G + SH1011	5 mM H_2SO_4 1 mL/min	10	Shodex application note – saccharides in lactic acid beverages and fermented milk

RID is utilized primarily for analytes that do not absorb in the UV-VIS, and is commonly applied in carbohydrate HPLC analysis, but less used in IC (Table 8).

2.1.9.4 Electrochemical detectors (ECD)

Electrochemical detectors are the best choice when the mobile phase is conductive and the analytes are electrochemically active (e.g., they can be oxidized or reduced by an electric current). Ions that can be electrochemically oxidized are more conveniently detectable (e.g., amines, azide, bromate, bromide, carbohydrates, chloride, chlorate, cyanide, iodate, iodide, nitrate, nitrite, sulfide, sulfite, thiocyanate, thiosulfate, tetrationate, etc.). The major advantages of ECD are sensitivity and specificity: they can offer limits of detection up to picogram/L levels, while being more selective than conductivity detectors, because only the compounds involved in the redox processes at the selected potential are detected (De Borba and Rohrer, 2007; Espinoza, 2020; Venard et al., 2020).

There are more approaches in electrochemical detection: amperometry (measuring current at a constant potential), pulsed amperometry (measuring current at constant potential pulses), coulometry (measuring current at a constant potential and integral conversion of the analyte) and voltammetry (measuring the current against potential in a defined range of potential).

Amperometric detection is the most popular of the ECD techniques; a fixed potential is applied between a working electrode (most often made of glassy carbon, gold, platinum or silver) and a reference electrode and then, an analyte, which will oxidize or reduce at that potential, yields an output current. Although only a small amount of the analyte is involved in the electron transfer process, the method is very sensitive.

In coulometric detection, 100% of the solute species is converted, which offers the advantages of no mobile-phase flow dependence on the signal and quantitation. The efficient use of an ECD is more complex, requiring detailed knowledge of the oxidation or reduction potential of the analyte(s) of interest.

In pulsed amperometric detection (PAD), analytes are detected by a redox process at the surface of an electrode, a rapidly repeating sequence of potentials being applied for removing the products of reactions from the electrode surface; in this way, the electrode surface is cleaned and activated, before each measurement is recorded. PAD is commonly used for the detection of electroactive species such as carbohydrates, sugar alcohols, aminoacids and sulfur species.

The ECD flow cell includes three electrodes: a working electrode (where the redox reaction occurs), an auxiliary electrode (through which the potential is applied and the current monitored) and a reference electrode (ensuring a fixed potential). When the electroactive species flows through a cell to which a potential is applied, this potential generates a background current from the eluent, which varies as an analyte undergoes oxidation or reduction. The magnitude of the resulted signal depends on the cell design, the nature of the analyte, its concentration and the applied voltage.

Mobile phases used with ECD have to be conductive; they must contain an adequate concentration of a fully ionized salt and should not undergo any electrochemical reaction in the flow cell. The salt concentration is not critical, as long as it is adequate for conduction (~0.05–0.1 M); at low levels, the background noise increases. The salt used should not interfere with the chromatography and should provide an appropriate pH for the separation and detection; if a gradient elution is required for a separation, there must be no major changes in the salt concentration through the gradient.

Table 9: Applications of IC analysis using amperometric detection.

Analytes/matrix	Column	Mobile phase	Separation time (min)	Reference
Biogenic amines (dopamine, tyramine, putresceine, cadaverine, histamine, serotonin agmantine, phenylethylamine, spermidine, spermine)	IonPac CS18	3–45 mM methanesulfonic acid/gradient 0.3 mL/min	42	De Borba and Rohrer (2007); Thermo Scientific/ Dionex – application note 182
Carbohydrates (arabinose, galactose, glucose, sucrose, fructose, lactose and maltose) in dairy products	CarboPac PA1	1 M CH_3COONa/0.2 M NaOH/H_2O/25 mM CH_3COONa (gradient) 0.25 mL/min	35	Hu and Rohrer (2020)
Carbohydrates (galactose, glucose, N-acetylgalactosamine, lactose, lactulose and epilactose) in milk whey	CarboPac PA1	10 mM NaOH + 2 mM $Ba(OAc)_2$ 1 mL/min	20	Cataldi et al. (2003)
Carbohydrates (sucrose, galactose, glucose, lactose and lactulose) in milk and dairy products	CarboPac PA20	10 mM KOH 0.008 mL/min	12	Christison et al. (2014)
Carbohydrates (arabinose, galactose, glucose, sucrose, fructose, lactose, isomaltulose, maltose) in food products	CarboPac PA20 (3 × 150 mm)	Water/0.1 M NaOH/ 0.2 M NaOH/0.6 M NaOH (gradient) 0.5 mL/min	30	Vennard et al. (2020)
Fermentable sugars (glucose, fructose, isomaltose, sucrose, maltose, maltotriose) in wort	CarboPac PA1	Water/0.5 M NaOH 1 mL/min	30	Thermo Scientific application note 46

Table 9 (continued)

Analytes/matrix	Column	Mobile phase	Separation time (min)	Reference
Fucose, fructose, saccharose and maltose oligosaccharides in wort	Metrosep Carb 2	100 mM NaOH + 25 mM CH_3COONa / 200 mM NaOH + 220 mM CH_3COONa 0.7 mL/min	60	Metrohm, IC application note P-084
Glycerol and ethanol in beer	IonPac ICE-AS6	100 mM $HClO_4$ 1 mL/min	12	Thermo Scientific application note 46
Iodide in milk products	IonPac AS11 Analytical, 4 × 250	50 mM HNO_3 1.5 mL/min	4	Dionex/ Thermo Scientific, Application Note 37
Lactose (residual) + galactose, fructose, sucrose and lactulose in lactose-free milk and dairy products	CarboPac PA20-Fast -4 µm	200 mM NaOH, 100 mM CH_3COONa, 1 M CH_3COONa 0.25 mL/min	15	Thielel and Jensen (2018)
Lactose (residual) + arabinose, fucose galactose, glucose, fructose, sucrose, alolactose, lactulose, epilactose and raffinose in lactose-free dairy products	CarboPac PA210	KOH/Eluent Generator 0.8 mL/min	12	Aggrawal and Rohrer (2018)
Lactose (residual) in lactose-free milk	Metrosep Carb 2–150/4.0	5 mM NaOH/2 mM CH_3COONa 0.8 mL/min	20	Metrohm, IC application note P-55
Lactose (residual) + galactose, glucose, sucrose, lactulose in lactose-free milk products	CarboPac PA20	200 mM NaOH, 100 mM CH_3COONa, 1 M CH_3COONa 0.5 mL/min	30	Perati et al. (2016)
Malto-oligosaccharides (glucose, maltose, maltotriose, maltotetraose, maltopentaose, maltohexaoze, maltoheptaose, maltooctaose, maltodecaose) in beer	CarboPac PA 100	Water/0.5 M NaOH/ 1 M CH_3COONa 1 mL/min	30	Thermo Scientific application note 46

Table 9 (continued)

Analytes/matrix	Column	Mobile phase	Separation time (min)	Reference
Organic acids (pyruvic, citric, malic, formic, lactic, acetic, succinic) in beer	IonPac ICE-AS6	0.8 mM heptafluorobutyric acid 0.8 mL/min	24	Thermo Scientific application note 46
Sulfite in wine	Shodex RS pack KC-811	12 mM H_3PO_4 1.2 mL/min	16	Shodex application note – KC-811
Sulfite in wine	IonPac ICE-AS1	20 mM methanesulfonic acid 0.2 mL/min	20	Chen et al. (2016)
Sulfite in wine	Metrosep Carb 2	0.3 M NaOH + 0.3 M CH_3COONa 0.5 mL/min	8	Espinosa (2020)

ECD requires a pulse-free flow (variations in the flow rate will generate supplementary noise), an efficient mobile-phase degassing (even tiny bubbles cause noise and loss of sensitivity) and a proper thermal stability (hence, a column oven should be used).

Despite being a demanding detection technique, ECD has found many applications in food analysis (Table 9), mainly because they are extremely selective and sensitive, while being very robust.

2.1.10 The waste reservoir

The last component of an IC system is the waste reservoir; usually it is a large glass or plastic bottle placed on the floor, below the IC system, which receives the eluent as it leaves the detector. Since this bottle has no sensor, it will not provide a feedback related to the filling level and, hence, it is important to ensure that it does not overflow, especially during long unattended runs for systems equipped with autosamplers.

2.1.11 The chromatographic data system

IC users need an appropriate software application, adapted to their work environment, to assist them control, organize, search, interpret and report chromatographic results. A chromatographic data system (CDS) is a specialized software running on a computer connected with the IC system, designed to receive and store the detector output, process chromatographic data (identify the chromatographic peaks, calculate peak area/height/width/asymmetry, discriminate between noise and peaks, perform calibrations, etc.), generate reports with the desired chromatographic information (peak heights, peak areas, sample identification, method variables, etc.), as well as control and diagnose the whole system. A continuous system monitoring over time ensures high analytical performance, allowing the user to diagnose and alter conditions, before maintenance is required (Dong et al., 2019; Mazzarese et al., 2019).

Remote access is a new option for certain CDSs and, hence, the system control becomes possible with any Wi-Fi-enabled tablet or mobile device connected to an instrument and loaded with a suitable control app. Such an option is of real interest when dealing frequently with large batches of samples running overnight, since the user can have a proper feedback whenever an unexpected event occurs, with the chance to resolve it in due time.

CDSs also allow a differentiated access to instruments, according to the hierarchy from the laboratory unit, where access rights are based on a unique log-on control, being differentiated for example, as follows.:

- for analytical user – can carry out sample analysis/process data/create methods and sequences/generate reports;
- for analytical supervisor – review and data approval, data queries, audit trail checks;
- for lab administrator – can add/configure instruments;
- for quality assurance – can review data/perform data queries;
- for CDS administrator – has full access rights.

There are different producers of IC systems in the market and each developed its own CDS, several cases being reviewed below.

MagIC Net CDS belongs to Metrohm; it manages the system control of their IC systems and also the service intervals. With a clear, intuitive graphical user interface, with a well-organized layout, database functions, it allows easy operation, followed by a proper data management, offering many possibilities for making easy and reliable determinations (MagIC Net, 2022). Empower 3^{TM} (Waters) and OpenLab (Agilent) can control also the IC systems produced by Metrohm based on dedicated drivers (Metrohm Press release 1, Metrohm Press release 2).

Chromeleon is the CDS controlling IC systems from Thermo Fisher Scientific (formerly Dionex), integrating system control and data handling. It has a very good functionality, with the most often needed features within view, while the areas of interest are easily accessible; the home screen shows overall system status, with tabs providing easy access to module functions (Chromeleon™ CDS Software). As main features, one can outline the following:

- uses a "plug-and-play" communication protocol via USB, providing easy configuration and safe connections;

- offers customizable reports and audit trail logs with details of instrument use, with a history file containing all manipulations;
- enables simpler troubleshooting and ensures good laboratory practice (GLP) compliance;
- stores results in secured databases, with effective query functions;
- has powerful network capabilities;
- monitors the consumables' status, identifying when the performance of certain components drifts outside recommended operating parameters and the component needs replacement;
- provides an efficient diagnosis, alerting potential system issues by a system wellness feature;
- provides also relevant information on the instrument's performance, tracking data and identifying changes in system performance over time;
- is easy to operate because of the wizards embedded in the software;
- offers a comprehensive assistance for the user by installation guides accompanied by videos and troubleshooting guides accessible through the interface.

Clarity Chromatography Software is a "third party" CDS developed by DataApex (a company focused only on chromatography software development), intended to provide a universal and more intuitive platform for most of the commercially available chromatographic systems; a list of currently controlled devices is available on the DataApex website (Clarity Chromatography Software). It represents an important step forward for the existing CDSs, because no producer provides a compatible platform for similar equipment. Clarity is a convenient CDS especially in laboratories in which different brands of chromatographic instruments are used, since the users are not in the situation to learn how to use more software. Clarity has a strong position in the CDS market for its intuitive use, excellent performance, cost-effectiveness and proficient technical support, with strong emphasis on new laboratory standards and practices and extensive customer support. Its features include data acquisition from multiple instruments and direct control of the chromatographic systems, making Clarity suitable for laboratories with high demands on efficiency, high sample throughput and GLP standards. A free demo for this CDS is available for download, so users can explore the available features (Clarity Demo).

2.2 IC systems' and columns' manufacturers

IC systems are available in various constructive variants: both dedicated system and modular ones, as well as portable systems are commercially available. Besides, process IC are specially designed for in-line automatic analysis of technological processes from industrial plants, serving as primary transducers in the automated control and regulation systems; they work on samples taken directly from technological flows, are integrated into installations and have a continuous working regime. Portable IC systems are important for on-site environmental applications, being less relevant in food analysis.

Some desktop IC systems are designed horizontally, requiring an important bench space (bench footprint). To save space, most instruments are built vertically, and comprise stackable modules. Although a vertical design involves a smaller footprint, it can cause problems related to hardware access for maintenance and troubleshooting.

There are several manufacturers of IC systems in the market, offering a wide range of configurations, adapted to the users' needs and budget. Relevant differences between them lie in sensitivity and flexibility (automation, software abilities, possibility to use in-line sample preparation devices). The most visible are:
- Metrohm – Herisau, Switzerland (https://www.metrohm.com/en/products/ion-chromatography.html);
- Thermo Fisher Scientific – a large US enterprise which incorporated numerous well-known brands, including the famous Dionex (https://www.thermofisher.com/ro/en/home/industrial/chromatography/ion-chromatography-ic.html);
- Shine – Shandong, China, a newer Chinese company (https://en.sheng-han.com);
- Sykam GmbH – Eresing, Germany (https://sykam.com/products/ion-chromatographs).

Besides, Antec Scientific-Zoeterwoude, The Netherlands (Antec Scientific) offers a dedicate carbohydrate analyzer ("Alexys"), which was shown to provide the requirements of AOAC 2018.16 method (Vennard et al., 2020). Alexys is, in fact, an adapted UHPLC system, equipped with a high-performance anion-exchange column, using pulsed amperometric detection (Antec Scientific), with the ability to separate soluble oligosaccharide in less than 35 min (Antec Scientific application note) from many food matrices.

Metrohm, Thermo Fisher Scientific and Shine offer also a wide range of IC columns; besides, one can consider:
- Shodex (SHOWA DENKO K.K., Tokyo, Japan) – offers IC columns for both anion and cation analysis (https://www.shodex.com/en/da1/07/#!);
- Macherey-Nagel – Macherey-Nagel AG Oensingen Switzerland (https://www.mn-net.com);
- Alltech™ – with IC columns under the brand Grace^R (https://grace.com);
- Vertisep – Vertical Chromatography Co.Ltd., Nonthaburi Thailand (http://www.vertichrom.com);
- Waters – Waters Corporation, USA (https://www.waters.com).

The simplest IC systems are the compact ones; they are most appropriate for routine analysis (such as those for quality control), easy to use, robust and with a small lab bench footprint, but usually they do not allow changes in the system's configuration (792 Basic IC, Metrohm; 930 Compact IC Flex, Metrohm; Eco IC, Metrohm; Thermo Scientific ICS-900).

Modular systems are more expensive but have the advantage of flexible configuration and are adapted to the user's needs (850 Professional IC Anion –MCS, Metrohm; Thermo Scientific ICS-3000; Thermo Scientific ICS-6000).

Irrespective of the IC system's type, in a buying decision for a new system, a good service and maintenance, including adequate training from the supplier, are essential for an analytical laboratory, for a proper activity.

References

792 Basic IC, Metrohm. (Accessed April 15, 2022, at https://partners.metrohm.com/GetDocument Public?action=get_dms_document&docid=623320)

850 Professional IC Anion – MCS, Metrohm (Accessed April 15, 2022, at https://www.metrohm. com/en/products/28502030)

930 Compact IC Flex, Compact ion chromatography system for routine analysis Metrohm. (Accessed April 15, 2022, at https://partners.metrohm.com/GetDocumentPublic?action=get_ dms_document&docid=1414490)

Acikara, Ö. B. (2013). Ion exchange chromatography and its applications. Column Chromatography, 10, 55744.

Aggrawal, M., & Rohrer, J. (2018). Determination of lactose in lactose-free dairy products using HPAE coupled with PAD and MS dual detection. Thermo Scientific, Application Note 72780. (Accessed October 19, 2022, at https://assets.thermofisher.com/TFS-Assets/CMD/Applica tion-Notes/an-72780-ic-ms-lactose-free-dairy-an72780-en.pdf).

Ahuja, S., Pohl, C. A., Avdalovic, N., & Srinivasan, K. (2021). Ion chromatography: Instrumentation, techniques and applications. Academic Press, Cambridge.

Al-Sanea, M. M., & Gamal, M. (2022). Critical analytical review: Rare and recent applications of refractive index detector in HPLC chromatographic drug analysis. Microchemical Journal, 178, 107339.

Antec Scientific application note – carbohydrates in food according to AOAC. (Accessed April 20, 2022, at https://antecscientific.com/download/18/carbohydrate-applications/8052/220_ 016-carbohydrates-in-food-according-to-aoac)

Antec Scientific. (Accessed April 20, 2022, at https://antecscientific.com/products/instruments/ alexys-analyzers)

Birkmann, J., Pasel, C., Luckas, M., & Bathen, D. (2018). UV spectroscopic properties of principal inorganic ionic species in natural waters. Water Practice & Technology, 13(4), 879–892.

Bruce, J. (2002). Analysis of anions in beer using ion chromatography. Journal of Automated Methods & Management in Chemistry, 24(4), 127–130.

Buchberger, W. W. (2001). Detection techniques in ion chromatography of inorganic ions. TrAC Trends in Analytical Chemistry, 20(6–7), 296–303.

Cataldi, T. R., Angelotti, M., D'Erchia, L., Altieri, G., & Di Renzo, G. C. (2003). Ion-exchange chromatographic analysis of soluble cations, anions and sugars in milk whey. European Food Research and Technology, 216(1), 75–82.

Cecchi, T. (2009). Ion-pair chromatography and related techniques. CRC Press, Boca Raton.

Chen, L., De Borba, B., & Rohrer, J. (2016). Determination of total and free sulfite in foods and beverages. Thermo Fisher Scientific application note 54. (Accessed August 11, 2021, at http://www.thermoscientific.es/content/dam/tfs/ATG/CMD/cmd-documents/sci-res/app/ chrom/ic/col/AN-54-IEX-Sulfite-Food-Beverage-AN70379-EN.pdf).

Christison, T., Verma, M., Kettle, A., Fisher, C., & Lopez, L. (2014). Determination of carbohydrates in beverages and milk products by HPAE-PAD and capillary HPAE-PAD. Thermo Scientific Poster Note PN71091_HPLC_2014_E_05/14S Thermo Fisher Scientific, Sunnyvale, CA, USA. (Accessed July 17, 2021, at https://tools.thermofisher.cn/content/sfs/posters/PN-71091-Carbohydrates-Beverages-Milk-HPAE-PAD-HPLC-2014-PN71091-EN.pdf).

Chromeleon™ CDS Software, Thermo Fisher Scientific. (Accessed April 15, 2022, at https://www.thermofisher.com/order/catalog/product/CHROMELEON7)

Clarity Chromatography Software. (Accessed April 15, 2022, at, https://www.dataapex.com/product/clarity-std)

Clarity Demo. (Accessed April 15, 2022, at https://www.dataapex.com/product/clarity-demo).

Colin, H., Guiochon, G., & Martin, M. (1986). Liquid chromatographic equipment. In H. Engelhardt (Ed) Practice of high performance liquid chromatography. Springer, Berlin, Heidelberg, 1–64.

Colin, H., Martin, M., & Guiochon, G. (1979). Extra-column effects in high-performance liquid chromatography: I. Theoretical study of the injection problem. Journal of Chromatography. A, 185, 79–95.

Cummins, P. M., Rochfort, K. D., & O'Connor, B. F. (2017). Ion-exchange chromatography: Basic principles and application. In: Walls, D., Loughran, S. (eds) Protein chromatography. Methods in Molecular Biology, vol 1485 Humana Press, New York, 209–223.

Dasgupta, P. K. (1992). Ion chromatography the state of the art. Analytical Chemistry, 64(15), 775A–783A.

De Borba, B. M., & Rohrer, J. S. (2007). Determination of biogenic amines in alcoholic beverages by ion chromatography with suppressed conductivity detection and integrated pulsed amperometric detection. Journal of Chromatography. A, 1155(1), 22–30.

Ding, X. J., Mou, S. F., Liu, K. N., Siriraks, A., & Riviello, J. (2000). Ion chromatography of heavy and transition metals by on-and post-column derivatizations. Analytica Chimica Acta, 407(1–2), 319–326.

Dionex/Thermo Scientific, Application Note 124 (2002). Determination of choline in dry milk and infant formula. (Accessed July 18, 2021, at https://assets.thermofisher.com/TFS-Assets/CMD/Application-Notes/4208-AN124_LPN1054-01.pdf).

Dionex/Thermo Scientific, Application Note 37 (2004). Determination of iodide in milk products. (Accessed July 17, 2021, at http://www.cromlab.es/Articulos/Columnas/HPLC/Thermo/Dionex/AS11/4128-AN37_24Jul95_LPN0702-03.pdf).

Dolan, J. (2014a). Column protection: Three easy steps. LC-GC North America, 32(12), 916–920.

Dolan, J. (2014b). Mobile-phase degassing: What, why and how. LC-GC North America, 32(7), 482–487.

Dolan, J. W. (2012). Avoiding refractive index detector problems. LC-GC North America, 30(12), 1032–1036.

Dong, M., Mazzarese, R., Zipfell, P., & Bird, S. (2019). Chromatography data systems: Perspectives, principles and trends. LCGC North America, 37(12), 852–865.

Dong, M. W. (2006). Modern HPLC for practicing scientists. John Wiley & Sons Inc. Hoboken, New Jersey.

Dong, M. W., & Wysocki, J. (2019). Ultraviolet detectors: Perspectives, principles and practices. LCGC North America, 37(10), 750–759.

Dorfner, K. (1991). 1.1 Introduction to ion exchange and ion exchangers. In Dorfner, K. (Ed.) Ion exchangers. de Gruyter, 7–188.

Eco IC, Metrohm. (Accessed April 15, 2022, at https://www.metrohm.com/en/products-overview/ion_chromatography/eco%20ic/29250020).

Espinosa, A. L. M. (2020). A simplified method to determine total sulfite content in food and beverages via ion chromatography. The Column, 16(2), 12–16.

Foley, J. P., & Dorsey, J. G. (1984). Clarification of the limit of detection in chromatography. Chromatographia, 18(9), 503–511.

Foley, R. C. L., & Haddad, P. R. (1986). Conductivity and indirect UV absorption detection of inorganic cations in non-suppressed ion chromatography using aromatic bases as eluents: I. Principles of operation. Journal of Chromatography. A, 366, 13–26.

Foster, K. L. (2005). Handbook of ion chromatography, 3rd ed. Wiley-VCH Verlag, Weinheim.

Fritz, J. S., & Gjerde, D. T. (2009) Ion chromatography, 4th ed. Wiley-VCH Verlag GmbH & Co. KGaA, Weinheim.

Fritz, J. S., & Gjerde, D. T. (2010). Discovery and early development of non-suppressed ion chromatography. Journal of Chromatographic Science, 48(7), 525–532.

Gaucheron, F., & Le Graet, Y. (2000). Determination of ammonium in milk and dairy products by ion chromatography. Journal of Chromatography. A, 893(1), 133–142.

Gjerde, D. T., Fritz, J. S., & Schmuckler, G. (1979). Anion chromatography with low-conductivity eluents. Journal of Chromatography. A, 186, 509–519.

Haddad, P. R., & Jackson, P. E. (1990). Ion chromatography. Elsevier, Amsterdam.

Haddad, P. R., Jackson, P. E., & Greenway, G. M. (1991). Ion chromatography: Principles and applications. Elsevier, Amsterdam.

Haidar Ahmad, I. A. (2017). Necessary analytical skills and knowledge for identifying, understanding and performing HPLC troubleshooting. Chromatographia, 80(5), 705–730.

Hu, J., & Rohrer, J. (2020). Determination of sugars in dairy products using HPAE-PAD. Thermo Scientific, Application Note 73341. (Accessed June 15, 2021, at https://assets.thermofisher.com/TFS-Assets/CMD/Application-Notes/an-73341-ic-hpae-pad-sugars-dairy-products-an73341-en.pdf).

Jackson, P. E., Thomas, D. H., Donovan, B., Pohl, C. A., & Kiser, R. E. (2001). New block-grafted anion exchanger for environmental water analysis by ion chromatography. Journal of Chromatography. A, 920(1–2), 51–60.

Klein, H., & Leubolt, R. (1993). Ion-exchange high-performance liquid chromatography in the brewing industry. Journal of Chromatography. A, 640(1–2), 259–270.

Klingenberg, A., & Seubert, A. (1998). Sulfoacylated polystyrene–divinylbenzene copolymers as resins for cation chromatography: Influence of capacity on resin selectivity. Journal of Chromatography. A, 804(1–2), 63–68.

Konczyk, J., Muntean, E., & Michalski, R. (2018). Simultaneous determination of chromate and common inorganic anions using suppressed ion chromatography. Chemistry, Environment, Biotechnology, 21, 11–13.

Kott, L. (2019). Ion chromatography: Method development. In G. K. Webster, L. Kott (Eds.) Chromatographic method development. Jenny Stanford Publishing, New York, 237–261.

Kraiczek, K. G., Bonjour, R., Salvadé, Y., & Zengerle, R. (2014). Highly flexible UV–VIS radiation sources and novel detection schemes for spectrophotometric HPLC detection. Analytical Chemistry, 86(2), 1146–1152.

Kriukova, A. I., Vladymyrova, I. M., Levashova, O. L., & Tishakova, T. S. (2019). Determination of amino acid composition in the harpagophytum procumbens root. Dhaka University Journal Of Pharmaceutical Sciences, 18(1), 85–91.

Kromidas, S. (Ed.) (2016). The HPLC expert: Possibilities and limitations of modern high performance liquid chromatography. John Wiley & Sons, Weinheim.

LaCourse, M. E., & LaCourse, W. R. (2017). General instrumentation in HPLC. In Fanali, S., Haddad, P. R., Poole, C., & Riekkola, M. L. (Eds.) Liquid chromatography, Chapter 17, 417–429. Elsevier, Amsterdam.

MagIC Net – Intelligent software for ion chromatography. (Accessed April 15, 2022, at https://www.metrohm.com/en-ae/products-overview/ion_chromatography/magic-net)

Marshall, M. R., Schmidt, R. H., & Walker, B. L. (2017). Ion chromatography for the food industry. In Fung, D. C. & Matthews, R. F. (eds.), Instrumental methods for quality assurance in foods. M. Dekker, New York, 39–66.

Marten, S. (2003). Ion chromatography determination of oxalic acid in beer. LC-GC Europe, Suppl.S, 12–13.

Mazzarese, R. P., Bird, S. M., Zipfell, P. J., & Dong, M. W. (2019). Chromatography data systems: Perspectives, principles, and trends. LC-GC North America, 37(12), 852–858.

McMaster, M. C. (2007). HPLC: A practical user's guide. John Wiley & Sons, Hoboken, New Jersey.

Metrohm IC Application Note C-105. Copper, zinc, iron(II) and manganese in wine by ion chromatography with post-column reaction and UV/VIS detection. (Accessed August 19, 2021, at https://partners.metrohm.com/GetDocumentPublic?action=get_dms_document&docid=694650).

Metrohm, IC application note CS-004, Determination of choline in baby milk powder. (Accessed July 17, 2021, at https://partners.metrohm.com/GetDocumentPublic?action=get_dms_document&docid=2259008).

Metrohm IC application note D-002, Anions and cations in beer; streamlining beverage analysis with ion chromatography. (Accessed August 4, 2021, at https://partners.metrohm.com/GetDocumentPublic?action=get_dms_document&docid=4864162).

Metrohm, IC application note P-55, Residual lactose in lactose-free milk. (Accessed July 17, 2021, at https://partners.metrohm.com/GetDocumentPublic?action=get_dms_document&docid=1975953.

Metrohm, IC application note P-084, Determination of glucose, fructose, saccharose and maltose oligosaccharides in wort applying pulsed amperometric detection after dose-in gradient elution. (Accessed July 17, 2021, at https://partners.metrohm.com/GetDocumentPublic?action=get_dms_document&docid=3461202.

Metrohm IC Application Note S–12. Determination of lactate, chloride, nitrate, sulfite and phosphate in wine. (Accessed August 4, 2021, at https://partners.metrohm.com/GetDocumentPublic?action=get_dms_document&docid=696334).

Metrohm IC Application Note S–281 (2009). Inorganic and organic anions in wine applying inline ultrafiltration. (Accessed May 17, 2022, at https://www.metrohm.com/content/dam/metrohm/shared/application-files/AN-S-281.pdf).

Metrohm IC application note S–162, Iodide in milk powder. (Accessed August 4, 2021, at https://partners.metrohm.com/GetDocumentPublic?action=get_dms_document&docid=696481).

Metrohm IC application note S-225, Sulfite besides standard anions in beer on the Metrosep A Supp 10. (Accessed August 4, 2021, at https://partners.metrohm.com/GetDocumentPublic?action=get_dms_document&docid=696036).

Metrohm IC application note S-287: Tap water analysis for anions and cations using Metrohm intelligent partial loop technique. (Accessed February 21, 2022, at https://partners.metrohm.com/GetDocumentPublic?action=get_dms_document&docid=873021).

Metrohm IC application note S–297, Iodide, thiocyanate and perchlorate in milk applying inline dialysis. (Accessed July 17, 2021, at https://partners.metrohm.com/GetDocumentPublic?action=get_dms_document&docid=983897).

Metrohm IC Application Note S–362. Organic acid anions in wine applying a low-pressure gradient. (Accessed May 17, 2022, at https://www.metrohm.com/content/dam/metrohm/shared/appli cation-files/AN-S-362.pdf).

Metrohm IC Application Note S-396 (2021). Assessing wine quality with IC Organic acid analysis using suppressed conductivity detection. (Accessed August 11, 2021, at https://partners.met rohm.com/GetDocumentPublic?action=get_dms_document&docid=4869637).

Metrohm Press Release 1 – Metrohm IC System "empowered" by Waters "Empower 3" software. (Accessed April 15, 2022, at https://www.metrohm.com/en-us/company/news/news-metrohm-ic-with-empower-3).

Metrohm Press Release 2 – Metrohm ion chromatography systems integrated in Agilent OpenLab CDS. (Accessed April 15, 2022, at https://www.metrohm.com/en-us/company/news/news-metrohm-ic-integrated-in-agilent-openlab-cds).

Meyer, V. R. (2013). Practical high-performance liquid chromatography. John Wiley & Sons, Chichester.

Michalski, R. (2003). Toxicity of bromate ions in drinking water and its determination using ion chromatography with post column derivatization. Polish Journal of Environmental Studies, 12, 6.

Michalski, R. (2009). Applications of ion chromatography for the determination of inorganic cations. Critical Reviews in Analytical Chemistry, 39(4), 230–250.

Michalski, R., & Łyko, A. (2010). Determination of bromate in water samples using post column derivatization method with triiodide. Journal of Environmental Science and Health Part A, 45 (10), 1275–1280.

Michalski, R., & Łyko, A. (2013). Ion chromatography with UV detection as sensitive method for bromate determination in bread. Asian Journal of Pharmaceutical Technology & Innovation, 01 (03), 1–8.

Michalski, R. (2014). Industrial applications of ion chromatography. Chemik, 68(5), 478–485.

Michalski, R. (2014a). Recent development and applications of ion chromatography. Current Chromatography, 1(2), 90–99.

Michalski, R., Muntean, E., & Łyko, A. (2021). Major inorganic ions in polish beers. Bulletin of University of Agricultural Sciences and Veterinary Medicine Cluj Napoca. Food Sciences and Technology, 78(1), 48–56.

Mohana Rangan, S., Krajmalnik-Brown, R., & Delgado, A. G. (2021). An IC method for simultaneous quantification of chromate, arsenate, selenate, perchlorate and other inorganic anions in environmental media. Environmental Engineering Science, 38(7), 626–634.

Morgan, N. Y., & Smith, P. D. (2010). HPLC detectors. Handbook of HPLC, 7, 207–231.

Nugent, K. D., & Dolan, J. W. (2019). Tools and techniques to extend LC column lifetimes. In Mant, C. T., & Hodges, R. S. (Eds.) High-performance liquid chromatography of peptides and proteins: Separation, analysis and conformation. CRC Press, Boca Raton, 31–35.

Parab, H., Ramkumar, J., Dudwadkar, A., & Kumar, S. D. (2021). Overview of ion chromatographic applications for the analysis of nuclear materials: Case studies. Reviews in Analytical Chemistry, 40(1), 204–219.

Paul, C., Steiner, F., & Dong, M. W. (2019). HPLC autosamplers: Perspectives, principles and practices. LC-GC North America, 37(8), 514–529.

Perati, P., De Borba, B., & Rohrer, J. (2016). Determination of lactose in lactose-free milk products by high-performance anion-exchange chromatography with pulsed amperometric detection. Thermo Fisher Scientific. Application Note 248. (Accessed July 17, 2021, at https://assets.ther mofisher.com/TFS-Assets/CMD/Application-Notes/AN-248-IC-Lactose-Milk-AN70236-EN.pdf).

Schaefer, J., Burmicz, J., & Palladino, D. (1989). Analysis using ion chromatography with electronic suppression. American Laboratory, 21(2), 70.

Scott, R. P. (1996). Chromatographic detectors: Design: Function and operation. CRC Press Boca Raton.

Shimadzu application news HPLC-022 (2019). The determination of 10 anions in EPA method 300.1 using Shimadzu high-resolution ion chromatography. (Accessed on 21 February,2021, at https://www.ssi.shimadzu.com/sites/ssi.shimadzu.com/files/ Products/literature/LC/022-EPA-Method-300-10-Anions.pdf).

Shodex application note – anions in beer. (Accessed July 17, 2021, at https://www.shodex.com/en/dc/03/08/13.html).

Shodex application note – anions in milk. (Accessed July 17, 2021, at https://www.shodex.com/en/dc/07/02/40.html).

Shodex application note – saccharides in lactic acid beverages and fermented milk. (Accessed July 17, 2021, at https://www.shodex.com/en/dc/03/02/32.html).

Shodex application note – KC-811. Sulfite in wine. (Accessed July 17, 2021, at https://www.shodex.com/en/dc/07/07/03.html).

Shodex application note SI-90 4E. Sulfite in wine. (Accessed July 17, 2021, at https://www.shodex.com/en/dc/07/05/14.html#!).

Shodex application note YS-50. Cations in red wine. (Accessed July 17, 2021, at https://www.shodex.com/en/dc/07/03/44.html).

Shoykhet, K., Broeckhoven, K., & Dong, M. W. (2019). Modern HPLC pumps: Perspectives, principles and practices. LC GC North America, 37(6), 374.

Skoog, D. A., West, D. M., Holler, F. J., & Crouch, S. R. (2013). Fundamentals of analytical chemistry. Cengage learning, Belmont.

Small, H., Stevens, T. S., & Bauman, W. C. (1975). Novel ion exchange chromatographic method using conductometric detection. Analytical Chemistry, 47(11), 1801–1809.

Snyder, L. R. (1980). Gradient elution. In Horváth, C. (Ed.) High-performance liquid chromatography: Advances and perspectives, Vol. 1, 207–316. Academic Press, New York.

Snyder, L. R., Kirkland, J. J., & Dolan, J. W. (2011). Introduction to modern liquid chromatography. John Wiley & Sons Inc, Hoboken, New Jersey.

Swartz, M. (2010). HPLC detectors: A brief review. Journal of Liquid Chromatography & Related Technologies, 33(9–12), 1130–1150.

Tanaka, K., & Haddad, P. R. (2000). Ion exclusion chromatography: Liquid chromatography. In Poole, C. F., Cooke, M., & Wilson, I. D. (Eds.) Encyclopedia of separation science. Academic Press, New York, 3193–3201.

Thermo Fisher Application note 140 (2001) Fast analysis of anions in drinking water by ion chromatography (former Dionex). (Accessed January 29, 2022, at https://assets.thermofisher.com/TFS-Assets/CMD/Application-Notes/4093-AN140_LPN1295.pdf).

Thermo Scientific application note 46 (2016). Ion chromatography: A versatile technique for the analysis of beer. (Accessed January 29, 2022, at https://tools.thermofisher.com/content/sfs/brochures/AN-46-IC-Beer-Analysis-AN71410-EN.pdf).

Thermo Scientific/Dionex – application note 21. Organic acids in wine. (Accessed January 29, 2022, at https://assets.thermofisher.com/TFS-Assets/CMD/Application-Notes/4118-AN21_LPN032025-02.pdf).

Thermo Scientific/Dionex – application note 182. Determination of biogenic amines in alcoholic beverages by ion chromatography with suppressed conductivity and integrated pulsed amperometric detections. (Accessed August 11, 2021, at http://www.cromlab.es/Articulos/Columnas/HPLC/Thermo/Dionex/CS18/56196-AN182_IC_BiogenicAmines_Alcohol_14Aug2007_LPN1888_02.pdf).

Thermo Scientific ICS-3000, Thermo Fisher Scientific. (Accessed January 29, 2022, at https://www.
  selectscience.net/products/thermo-scientific-ics-3000-high-performance-ion-chromatography
  -system/?prodID=1067).
Thermo Scientific ICS-900 – Thermo Fisher Scientific. (Accessed January 29, 2022, at https://www.
  selectscience.net/products/thermo-scientific-ics-900-starter-line-ic-system/?prodID=103857)
Thielel, K. J. F., & Jensen, D. (2018). Fast determination of lactose in dairy products. Thermo Fisher
  Scientific. Application Note 72633. (Accessed January 29, 2022, at https://assets.thermo
  fisher.com/TFS-Assets/CMD/Application-Notes/can-72633-ic-lactose-dairy-products-
  can72633-en.pdf)
US EPA Method 218.7 (2011). Determination of hexavalent chromium in drinking water by ion
  chromatography with post-column derivatization and UV-visible spectroscopic detection,
  United States Environmental Protection Agency. Office of Groundwater and Drinking Water,
  USEPA, Cincinnati. (Accessed February 21, 2022, at https://nepis.epa.gov).
Vennard, T. R., Ruosch, A. J., Wejrowski, S. M., & Ellingson, D. J. (2020). Sugar profile method by
  high-performance anion-exchange chromatography with pulsed amperometric detection in
  food, dietary supplements, pet food and animal feeds: First action 2018.16. Journal of AOAC
  International, 103(1), 89–102.
Vial, J., & Jardy, A. (1999). Experimental comparison of the different approaches to estimate LOD
  and LOQ of an HPLC method. Analytical Chemistry, 71(14), 2672–2677.
Walters, R. R. (1983). Column fittings without ferrules for liquid chromatography. Analytical
  Chemistry, 55(3), 591–592.
Weiss, J. (2016). Handbook of ion chromatography. John Wiley & Sons Hoboken, New Jersey.
Weiss, J., & Jensen, D. (2003). Modern stationary phases for ion chromatography. Analytical and
  Bioanalytical Chemistry, 375(1), 81–98.
Wouters, S., Haddad, P. R., & Eeltink, S. (2017). System design and emerging hardware technology
  for ion chromatography. Chromatographia, 80(5), 689–704.
Zerbinati, O., Balduzzi, F., & Dell'Oro, V. (2000). Determination of lithium in wines by ion
  chromatography. Journal of Chromatography. A, 881(1–2), 645–650.

3 IC applications in food analysis, quality control, food safety and food authentication

3.1 Water

Water is the most important food, being in the meantime a critical resource for food industry; consumed as it is or as a component in a wide variety of food products, water is essential to life. In the food industry, water is used not only as an ingredient but also as a processing aid (for dissolving, diluting, cleaning operations, rinsing, dispersing, conveying, heating, cooling, blanketing, separating, steam generation, etc.), and ending as wastewater. Mineral waters are a particular type of water, originating from under the ground. "Natural mineral water means microbiologically wholesome water [. . .] originating in an underground water table or deposit, and emerging from a spring tapped at one or more natural or bore exits" (Directive 2009/54/EC). Most mineral waters are characterized by a high content of dissolved ions, the bio-availability of magnesium and calcium being of a wider interest. Despite the consumption of mineral water has been systematically rising in the highly industrialized countries, it is not meant to be the main source of daily water intake (Boyd, 2019; De la Guardia and Garrigues, 2015).

Regardless of its type, water quality is an important issue to be monitored. When water is an ingredient, its quality can affect the properties of food products, including their texture, shelf stability, appearance, aroma and flavor; as a processing aid, an improper water quality can have negative effects on equipment or even on the technological process (e.g., hard water may cause deposits on equipment surfaces, pipes, valves/ reduce water's ability to dissolve and disperse food ingredients/reduce foaming of soaps and rinsing effectiveness in cleaning applications, etc.); as wastewater, improper loading with certain compounds can have negative impact on the environment. Hence, an improper water quality may lead to food safety and product quality issues, and cause problems for equipment operation and maintenance (Bhagwat, 2019).

All water quality issues are related to components contained in the water matrix, either dissolved or dispersed: water contains many inorganic and organic compounds originating from water sources, treatment processes and contact with different materials; some of them, such as calcium, potassium, magnesium and sodium, are essential to humans, since the human body does not synthesize them. Besides, drinking water can also contain many other trace essential elements, such as chromium, cobalt, iron, copper, manganese, selenium and zinc. The most common salts dissolved in water are chlorides, sulfates and hydrogen carbonates of sodium, calcium and

https://doi.org/10.1515/9783110644401-003

magnesium. To express water quality, total hardness,[8] temporary hardness[9] (carbonate hardness) and residual hardness[10] are the frequently used parameters (Boyd, 2019; Rubenowitz-Lundin and Hiscock, 2013).

Being so important for human health, water quality is regulated. In the European Union, the Drinking Water Directive (Directive 2020/2184) is a well-known document, its last version being effective from 12th January 2021, reinforcing the water quality standards, imposing tighter limits for certain analytes (e.g., bromate, chlorate and chlorite) and requiring Member States to transpose them into their national legislation within two years. To ensure compliance with legal regulations, IC is a reference method of choice for the determination of numerous analytes from water (Michalski, 2006), replacing many of the current classical "wet" chemical methods in most accredited and research laboratories. In fact, the determination of inorganic ions in water is the most widely used applications of IC, allowing a fast and convenient determination of both anions and cations. Due to its major advantages, especially in the determination of anions, many international standards have regulatory methods of analysis based on IC (e.g., ASTM D4327-17, D5257-17, D6581-18, D6919-17, D6994-15; ISO 10304-1/2007, 10304-3/1007, 10304-4/1997, 11206/2011, 14911/1998, 15061/2001; US EPA Methods 218.7/2011, 300/1993, 302/2009, 314/1999, 314.1/2005, 314.2/2008, 317/2000, 321.8/1997, 326/2002, 331/2005, 332/2005, 557/2009, 9056A/2007). IC applications for water analysis are summarized in Table 10.

IC analysis of inorganic anions from water samples can be accomplished conveniently using IC, either by isocratic separations with carbonate/bicarbonate mobile phases or by using electrolytically generated hydroxide eluents (by reagent-free IC, this option enabling both isocratic and gradient elution). From anions, nitrite and nitrate are among the most undesirable components of the water matrix. Nitrite can interact with hemoglobin, forming methemoglobin and disturbing the ability of blood to transport oxygen, while nitrate can react with amines, producing potential carcinogens, mutagenic and/or teratogenic N-nitrosamines (Parvizishad et al., 2017; Schullehner et al., 2017). Columns such as Metrosep A Supp 5, Metrosep A Supp 7, Metrosep A Supp 17, Shodex IC SI-52 4E, Dionex IonPac AS1, can be successfully used for compliance monitoring of inorganic anions in drinking water and wastewater in accordance with US EPA Methods 300.0 and 300.1 (Konczyk et al., 2019; Metrohm application notes S-287 and AN-S-353; Mohana Rangan et al., 2021; Shimadzu application news HPLC-022). Nitrite and nitrate can be determined, besides bromide, using direct UV detection at 210 nm; this method is advantageous in some

---

8 The total hardness of water represents the sum of the molar concentrations of calcium and magnesium ions.
9 The temporary hardness is that part of the hardness that can be removed by boiling.
10 The residual hardness is the difference between the former two parameters, which is due to the sulfates and chlorides of calcium and magnesium that cannot be precipitated out by boiling.

Table 10: Summary of IC applications for water analysis.

Analytes/matrix	Column	Mobile phase	Detection	Sample preparation	Separation time (min)	Reference
Fluoride, chloride, nitrate, sulfate	Metrosep A Supp 5 – 250/4	3.2 mM Na$_2$CO$_3$/ 1 mM NaHCO$_3$ 0.7 mL/min	Suppressed conductivity	Filtration	14	Metrohm IC application note S-287
Fluoride, chloride, nitrite, nitrate, selenate, arsenate, perchlorate, chromate	Metrosep A Supp 7	10.8 mM Na$_2$CO$_3$/ 35% acetonitrile 0.8 mL/min	Suppressed conductivity	Filtration	20	Mohana Rangan et al. (2021).
Chloride, fluoride, bromide, nitrate, nitrite, phosphate, sulfate, carbonate.	Metrosep A Supp 17 – 150/4	Na$_2$CO$_3$/NaHCO$_3$	Suppressed conductivity	Filtration	18	Metrohm application note AN-S-353
Fluoride, chlorite, bromate, chloride, nitrite, bromide, chlorate, nitrate, phosphate, sulfate	Shodex IC SI-52 4E, 250/4	5.4 mM Na$_2$CO$_3$ 0.8 mL/min	Suppressed conductivity	Filtration	20	Shimadzu application news HPLC-022
Fluoride, chloride, bromide, nitrate, phosphate, sulfate, chromate	Dionex IonPac AS1	5 mM Na$_2$CO$_3$/1 mM NaHCO$_3$ 1 mL/min	Suppressed conductivity	Filtration	14	Konczyk et al, (2018)
Fluoride, bromate, chloride, bromide nitrate, sulfate	Dionex IonPac AS23, 250/4	4.2 mM Na$_2$CO$_3$/ 1 mM NaHCO$_3$ 1.1 mL/min	Conductivity	Filtration	25	Djam et al. (2019)
Fluoride, chloride, nitrite, nitrate, phosphate, sulfate	Dionex IonPac AS22-Fast (2x150)	4.5 mM Na$_2$CO$_3$/ 1.4 mM NaHCO$_3$, 0.5 mL/min	Suppressed conductivity	Filtration	<5	Thermo Fisher App. note 120

(continued)

Table 10 (continued)

Analytes/matrix	Column	Mobile phase	Detection	Sample preparation	Separation time (min)	Reference
Nitrite, bromide, nitrate	Dionex IonPac AS9	1.8 mM Na_2CO_3/1.7 mM $NaHCO_3$, 2 mL/min	UV (210 nm)			Thermo Fisher App. note 132
Fluoride, chloride, nitrite, nitrate, phosphate, sulfate	Dionex IonPac AS14A (3 × 150)	8 mM Na_2CO_3/1 mM $NaHCO_3$, 0.8 mL/min	Suppressed conductivity	Filtration	6	Thermo Fisher App. note 140
Disinfection by-products (monobromo-, dibromo-, monochloro-, dichloro-, trichloro-, chlorodifluoro- and trifluoroacetic acids besides fluoride, chloride, nitrate and sulfate in drinking water	Dionex AS11HC, 250/2 or AS16, 250/2	10–100 mM KOH – gradient 0.3 mL/min 2.5–20 mM KOH – gradient 0.3 mL/min	Suppressed conductivity	Acidification, solid phase extraction	35 25	Barron, L. and Paull, B. (2004)
Perchlorate, down to 1.5 µg/L, in drinking and surface water, swimming pool water	Dionex IonPac AS20	35 mM KOH (eluent generator Dionex RFC-30) 0.25 mL/min	Suppressed conductivity	Cation exchange/filtration	30	Seiler et al. (2016)
Perchlorate	Concise AN1HS 50/3	6 mM Na_2CO_3 0.7 mL/min	Suppressed conductivity + MS	Filtration	10	Shimadzu application news HPLC-035

Analytes	Column	Eluent / flow	Detection	Sample preparation	Runtime	Reference
Fluoride, chloride, nitrate, phosphate, sulfate	Metrosep A Supp 5, 150/4	3.6 mM Na_2CO_3/1 mM $NaHCO_3$ 0.7 mL/min	Suppressed conductivity	Degassing/dilution/filtration	n.a.	Konczik et al. (2019)
Potassium, sodium, magnesium, calcium	Metrosep C6 150/4	1.7 mM pyridine-2,6-dicarboxylic acid / 1.7 mM HNO_3 0.9 mL/min	Nonsuppressed conductivity	Degassing/dilution/filtration	n.a.	
Lithium, sodium, potassium, calcium, magnesium	Metrosep C 4, 100/2	2 mM HNO_3/1.2 mM dipicolinic acid 0.5 mL/min	Conductivity		6	Metrohm IC application note AN-C-147
Lithium, sodium, ammonium, potassium, calcium, magnesium, strontium, barium	Metrosep C 4 – 150/4	2 mM HNO_3/2 mM dipicolinic acid 0.9 mL/min	Conductivity		20	Metrohm IC application note C-135
Lithium, sodium, potassium, calcium, magnesium	Metrosep C 4	1.7 mM HNO_3/0.7 mM dipicolinic acid 0.5 mL/min	Conductivity	Acidification to pH = 3	22	Metrohm IC application notes AN-C-115, C-133
Sodium, potassium, calcium, magnesium	Metrosep C 6, 250/4	1.7 mM HNO_3 0.9 mL/min	Conductivity		12	Metrohm IC application note AN-C-154
Sodium, potassium, calcium, magnesium	Metrosep C 6	6.75 mM HNO_3 0.25 mL/min	Conductivity		12	Metrohm IC application note AN-C-174
Lithium, sodium, potassium, calcium, magnesium	Dionex IonPac CS16-4µm	30 mM methanesulfonic acid 0.9 mL/min	Suppressed conductivity	Filtration	22	Thermo Fisher Application note 204

cases since it is free from most ionic interferences due to the specificity of UV detection (Thermo Fisher Application note 132).

Among anions from certain water matrices, the by-products resulting from water disinfection are of particular interest due to their effect on human health. Drinking water disinfection is a must to destroy bacteria, parasites or viruses that can cause certain diseases in humans. Disinfection is carried out by chlorination or ozonation. Unfortunately, in both cases, the disinfection treatment process produces byproducts that are themselves health hazards: chlorination can lead to several hazardous compounds, such as trihalomethanes or haloacetic acids, while ozonation, despite being more advantageous, can also lead to harmful byproducts, such as bromate, with proven toxicity and suspected mutagenicity, which can increase the risk of cancer (Michalski and Lyko, 2013). Such analytes can be determined from drinking water using the Dionex IonPac AS23 column, with a carbonate/bicarbonate eluent for the determination of chlorite, bromate and chlorate, together with common inorganic anions, as well as with Shodex IC Si-52 4E column (Figure 43).

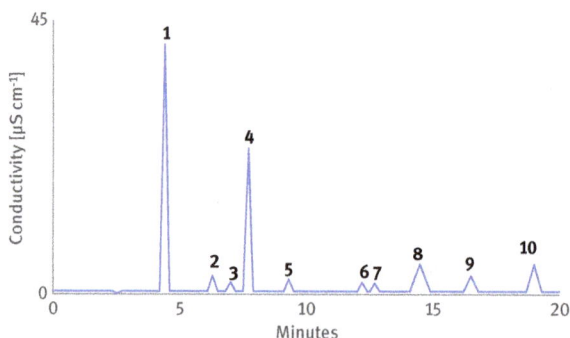

Figure 43: Separation of anions from water using a Shodex IC Si-52 4E column, with 5.4 mM Na$_2$CO$_3$, 0.8 mL/min. Peak ID: 1, fluoride; 2, chlorite; 3, bromate; 4, chloride; 5, nitrite; 6, bromide; 7, chlorate; 8, nitrate; 9, phosphate; 10, sulfate (Shimadzu application news HPLC-022).

Another anion of concern in the last few years is perchlorate; it has been found in areas where rocket fuel, ammunition, matches or pyrotechnic articles are manufactured, tested or used, but sources for the contamination with this anion also include certain chemical fertilizers and air bag inflators (Trumpolt et al., 2005). Perchlorate presents a health risk for humans because it can cause dysfunctions of the thyroid gland, decreasing its ability to produce thyroid hormones (Charnley, 2008; Seiler et al., 2016). Several IC techniques were standardized by US EPA for the determination of perchlorate (US EPA 314.0, 2005; US EPA 314.1, 2005; US EPA 314.2, 2008), all of them using suppressed conductivity detection. For determination in the low ng/L range, US EPA standardized two more methods, US EPA 331.0 and US EPA 332, both based on suppressed anion chromatography, coupled with mass spectrometry. The anion-exchange column Dionex IonPac AS20 can be used for the analysis of trace amounts of perchlorate using suppressed conductivity detection in the presence of much larger concentrations of chloride, sulfate and carbonate.

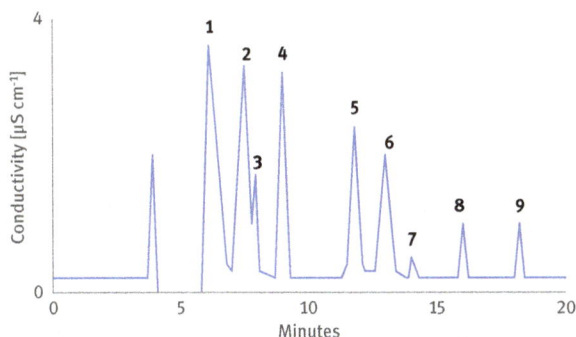

Figure 44: Separation of anions from water with a Metrosep A Supp 7 column, with 10.8 mM Na_2CO_3 and 35% acetonitrile. Peak ID: 1, fluoride; 2, chloride; 3, nitrite; 4, nitrate; 5, sulfate; 6, selenate; 7, arsenate; 8, perchlorate; 9, chromate (Mohana Rangan et al., 2021).

Chromate is the last anion of concern discussed here. It contains the toxic, mutagenic and carcinogenic chromium (VI) (Salnikow and Zhitkovich, 2008), for which the World Health Organization (WHO) suggests a maximum allowable limit of 50 µg/L in drinking water (WHO, 2020). The presence of chromate can be determined using a high-capacity anion-exchange separator column, postcolumn derivatization with 1,5-diphenylcarbazide and UV-Vis detection at 530 nm (EPA Method 218.7). The presence of chromate, besides other anions, can also be determined using a Dionex IonPac AS1 column (Konczyk et al., 2018) or a Metrosep A Supp 7 column (Figure 44) with suppressed conductivity (Mohana Rangan et al., 2021).

Cation analysis is another important IC application in the field of drinking water analysis, allowing the simultaneous determination of alkali and alkaline-earth metals. Cation analysis is of special importance when dealing with mineral waters, since they are important sources of micro- and macroelements for humans. Their composition is highly dependent on both the geochemical features and the anthropogenic pollution (Diduch et al., 2011; Jung, 2001). Using high capacity cation columns, such as Metrosep C6 – 250/4.0, and a strong eluent, such as nitric acid, cations from drinking water can be determined within a short runtime (Figure 45). Parallel anion and cation determinations from water samples are also possible using specially-designed systems (Fa et al., 2018; Metrohm application note C–133; Metrohm application note S-287; Muntean et al,. 2008).

Last but not the least, sample preparation for water analysis is remarkably simple. Drinking water, which is usually particle-free and not turbid, can be analyzed directly, while for turbid samples and samples containing particles, filtration is recommended

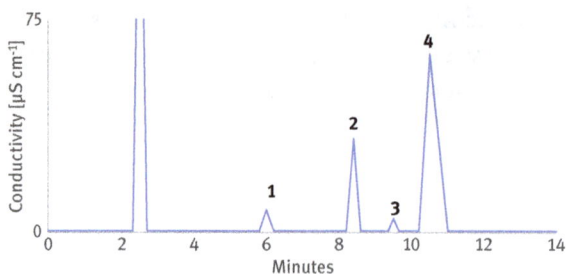

Figure 45: Separation of cations from water with a Metrosep C6 column, with 1.7 mM HNO_3, 0.9 mL/min. Peak ID: 1, sodium; 2, magnesium; 3, potassium; 4, calcium (Metrohm application note AN-C-154).

to prevent clogging of the system and to improve the column lifetime. In cation analysis, the samples have to be acidified by the addition of nitric acid up to a pH between 2.5 and 3.5 in order to obtain reproducible results. Samples have to be stored in plastic vessels.

References

ASTM D4327-17 Standard test method for anions in water by suppressed IC. ASTM International, West Conshohocken, PA. (Accessed February 21, 2022, at https://www.astm.org/Standards/D4327.htm).

ASTM D5257-17 Standard test method for dissolved hexavalent chromium in water by IC. ASTM International, West Conshohocken, PA. (Accessed February 21, 2022, at https://www.astm.org/Standards/D5257.htm).

ASTM D6581-18 Standard test methods for bromate, bromide, chlorate and chlorite in drinking water by suppressed ion chromatography. ASTM International, West Conshohocken, PA. (Accessed February 21, 2022, at https://www.astm.org/Standards/D6581.htm).

ASTM D6919-17 (2009) Standard test method for determination of dissolved alkali and alkaline earth cations and ammonium in water and wastewater by IC. ASTM International, West Conshohocken, PA. (Accessed February 21, 2022 at https://www.astm.org/Standards/D6919.htm).

ASTM D6994-15 Standard test method for determination of metal cyanide complexes in wastewater, surface water, groundwater and drinking water using anion exchange chromatography with UV detection. ASTM International, West Conshohocken, PA. (Accessed February 21, 2022, at https://www.astm.org/Standards/D6994.htm).

Barron, L., & Paull, B. (2004). Determination of haloacetic acids in drinking water using suppressed micro-bore ion chromatography with solid phase extraction. Analytica Chimica Acta, 522(2), 153–161.

Bhagwat, V. R. (2019). Safety of water used in food production. In Food safety and human health. Academic Press, 219–247.

Boyd, C. E. (2019). Water quality: an introduction. Springer Nature.

Charnley, G. (2008). Perchlorate: overview of risks and regulation. Food and Chemical Toxicology, 46(7), 2307–2315.

De la Guardia, M., & Garrigues, S. (2015). Handbook of mineral elements in food. John Wiley & Sons.

Diduch, M., Polkowska, Ż., & Namieśnik, J. (2011). Chemical quality of bottled waters: a review. Journal of Food Science, 76(9), R178–R196.

Djam, S., Najafi, M., Ahmadi, S. H., & Shoeibi, S. (2019). Determination of bromate in bottled water marketed in Iran by ion chromatography. Journal of Chemical Metrology, 13(2), 47–52.

Directive 2009/54/EC of the European Parliament and of the Council, Official Journal of the European Union of 18 June 2009 on the exploitation and marketing of natural mineral waters. (Accessed February 5, 2022, at https://eur-lex.europa.eu/legal-content/EN/TXT/PDF/?uri=CELEX:32009L0054&qid=1644140832670&from=EN).

Directive 2020/2184 of the European Parliament and of the Council, of 16 December 2020 on the quality of water intended for human consumption. (Accessed February 5, 2022, at https://eur-lex.europa.eu/legal-content/EN/TXT/PDF/?uri=CELEX:32020L2184&from=EN).

Fa, Y., Yu, Y., Li, F., Du, F., Liang, X., & Liu, H. (2018). Simultaneous detection of anions and cations in mineral water by two dimensional ion chromatography. Journal of chromatography. A, 1554, 123–127.

ISO 10304-1 (2007) Water quality – Determination of dissolved anions by liquid chromatography of ions – Part 1: Determination of bromide, chloride, fluoride, nitrate, nitrite, phosphate and sulfate. ISO, Geneva, Switzerland. (Accessed February 5, 2022, at https://www.iso.org/standard/46004.html).

ISO 10304-1:2007/COR 1 (2010) Water quality – Determination of dissolved anions by liquid chromatography of ions – Part 1: Determination of bromide, chloride, fluoride, nitrate, nitrite, phosphate and sulfate – Technical Corrigendum 1 ISO, Geneva, Switzerland. (Accessed February 5, 2022, at https://www.iso.org/standard/56509.html).

ISO 10304-3 (1997) Water quality – Determination of dissolved anions by liquid chromatography of ions – Part 3: Determination of chromate, iodide, sulfite, thiocyanate and thiosulfate. ISO, Geneva, Switzerland. (Accessed February 5, 2022, at https://www.iso.org/standard/20651.html).

ISO 10304-4 (1997). Water quality- Determination of dissolved anions by liquid chromatography of ions – Part 4: Determination of chlorate, chloride and chlorite in water with low contamination ISO, Geneva, Switzerland. (Accessed February 5, 2022, at https://www.iso.org/standard/22573.html).

ISO 11206 (2011). Water quality – Determination of dissolved bromate – method using IC and post column reaction. ISO, Geneva, Switzerland. (Accessed February 5, 2022, at https://www.iso.org/standard/50243.html).

ISO 14911 (1998). Water quality: determination of dissolved lithium, sodium, ammonium, potassium, manganese, magnesium, calcium, strontium and barium using IC – method for water and wastewater. ISO, Geneva, Switzerland, (Accessed February 5, 2022, at https://www.iso.org/standard/25591.html).

ISO 15061 (2001) Water quality – Determination of dissolved bromate – Method by liquid chromatography of ions. ISO, Geneva, Switzerland. (Accessed February 5, 2022, at https://www.iso.org/standard/25863.html).

Jung, M. C. (2001). Heavy metal contamination of soils and waters in and around the Imcheon Au–Ag mine, Korea. Applied Geochemistry, 16(11–12), 1369–1375.

Konczyk, J., Muntean, E., & Michalski, R. (2018). Simultaneous determination of chromate and common inorganic anions using suppressed ion chromatography. Chemistry, Environment, Biotechnology, 21, 11–13.

Kończyk, J., Muntean, E., Gega, J., Frymus, A., & Michalski, R. (2019). Major inorganic anions and cations in selected European bottled waters. Journal of Geochemical Exploration, 197, 27–36.

Metrohm application note AN-C-115: Five cations in tap water. (Accessed February 5, 2022, at https://partners.metrohm.com/GetDocumentPublic?action=get_dms_document&docid= 695415).

Metrohm application note AN-C-147: Fast analysis of cations in tap water using Metrosep C 4-100/ 2.0. (Accessed February 5, 2022, at https://partners.metrohm.com/GetDocumentPublic?ac tion=get_dms_document&docid=1452576).

Metrohm application note AN-C-154: Fast IC: Cations in drinking water on a high-capacity column in eleven minutes. (Accessed February 5, 2022, at https://partners.metrohm.com/GetDocument Public?action=get_dms_document&docid=1979772).

Metrohm application note AN-C-174: Rapid determination of cations in drinking water on a microbore separation column. (Accessed February 5, 2022, at https://partners.metrohm.com/ GetDocumentPublic?action=get_dms_document&docid=2667483).

Metrohm application note AN-S-353 Routine drinking water analysis – Robust analysis of major anions by ion chromatography. (Accessed February 21, 2022, at https://partners.metrohm. com/GetDocumentPublic?action=get_dms_document&docid=2595385).

Metrohm application note C–133: Tap water analysis for anions and cations using Metrohm intelligent Partial Loop Technique (MiPT). (Accessed February 5, 2022, https://partners.met rohm.com/GetDocumentPublic?action=get_dms_document&docid=858285).

Metrohm application note C-135: Cations in drinking water using Metrosep C 4-150/4.0 column according to ISO 14911. (Accessed February 5, 2022, at https://partners.metrohm.com/GetDo cumentPublic?action=get_dms_document&docid=983894).

Metrohm application note S-287: Tap water analysis for anions and cations using Metrohm intelligent Partial Loop Technique (MiPT) (Accessed February 5, 2022, at https://partners.met rohm.com/GetDocumentPublic?action=get_dms_document&docid=873021).

Michalski, R. (2006). Ion chromatography as a reference method for determination of inorganic ions in water and wastewater. Critical Reviews in Analytical Chemistry, 36(2), 107–127.

Michalski, R., & Łyko, A. (2013). Bromate determination: state of the art. Critical Reviews in Analytical Chemistry, 43(2), 100–122.

Mohana Rangan, S., Krajmalnik-Brown, R., & Delgado, A. G. (2021). An IC method for simultaneous quantification of chromate, arsenate, selenate, perchlorate and other inorganic anions in environmental media. Environmental Engineering Science, 38 (7), 626–634.

Muntean, E., Mihăiescu, T., Muntean, N., & Mihăiescu, R. (2008). Simultaneous ion chromatographic determination of anions and cations in surface waters from Fizes Valley. Chemické Listy, 102, s265–s1309.

Rubenowitz-Lundin, E., & Hiscock, K. M. (2013). Water hardness and health effects. In Essentials of Medical Geology (pp. 337–350). Springer, Dordrecht.

Parvizishad, M., Dalvand, A., Mahvi, A. H., & Goodarzi, F. (2017). A review of adverse effects and benefits of nitrate and nitrite in drinking water and food on human health. Health Scope, 6(3), e14164.

Salnikow, K., & Zhitkovich, A. (2008). Genetic and epigenetic mechanisms in metal carcinogenesis and cocarcinogenesis: nickel, arsenic and chromium. Chemical Research in Toxicology, 21(1),28–44.

Seiler, M. A., Jensen, D., Neist, U., Deister, U. K., & Schmitz, F. (2016). Validation data for the determination of perchlorate in water using ion chromatography with suppressed conductivity detection. Environmental Sciences Europe, 28(1),1–9.

Schullehner, J., Stayner, L., & Hansen, B. (2017). Nitrate, nitrite and ammonium variability in drinking water distribution systems. International Journal of Environmental Research and Public Health, 14(3), 276.

Shimadzu application news HPLC-022 (2019). The determination of 10 anions in EPA method 300.1 using Shimadzu high-resolution ion chromatography. (Accessed February 21, 2022, at https://www.ssi.shimadzu.com/sites/ssi.shimadzu.com/files/Products/literature/LC/022-EPA-Method-300-10-Anions.pdf).

Shimadzu application news HPLC-035 (2020). Perchlorate quantitation in drinking water using suppressed anion chromatography coupled with single quadrupole MS. (Accessed February 21, 2022, at https://www.ssi.shimadzu.com/sites/ssi.shimadzu.com/files/Industry/Literature/HPLC-035-Perchlorate-Quantitation-Drinking-Water.pdf).

Thermo Fisher Application note 120 (2016) Municipal drinking water analysis by fast IC. (Accessed February 21, 2022, at https://assets.thermofisher.com/TFS-Assets/CMD/Application-Notes/AB-120-IC-Municipal-Drinking-Water-Fast-IC-AB71940-EN.pdf).

Thermo Fisher Application note 132 (1991) Determination of nitrite and nitrate in drinking water using ion chromatography with direct UV detection (former Dionex). (Accessed February 21, 2022, at https://assets.thermofisher.com/TFS-Assets/CMD/Application-Notes/4189-AU132_Apr91_LPN034527.pdf).

Thermo Fisher Application note 140 (2001) Fast analysis of anions in drinking water by ion chromatography (Former Dionex). (Accessed February 21, 2022, at https://assets.thermofisher.com/TFS-Assets/CMD/Application-Notes/4093-AN140_LPN1295.pdf).

Thermo Fisher Application note 204 (2016) Analysis of environmental waters for cations and ammonium using a compact ion chromatography system. (Accessed February 21, 2022, at https://assets.thermofisher.com/TFS-Assets/CMD/Application-Notes/AU-204-IC-Cations-Ammonium-Water-AU72033-EN.pdf).

Trumpolt, C. W., Crain, M., Cullison, G. D., Flanagan, S. J., Siegel, L., & Lathrop, S. (2005). Perchlorate: sources, uses and occurrences in the environment. Remediation Journal: The Journal of Environmental Cleanup Costs, Technologies & Techniques, 16(1), 65–89.

US EPA Method 218.7 (2011). Determination of hexavalent chromium in drinking water by ion chromatography with postcolumn derivatization and UV-visible spectroscopic detection, United States Environmental Protection Agency. Office of Groundwater and Drinking Water, USEPA, Cincinnati. (Accessed February 21, 2022, at https://nepis.epa.gov).

US EPA Method 300 (1993). Determination of inorganic anions by IC, United States Environmental Protection Agency. Office of Groundwater and Drinking Water, USEPA, Cincinnati. (Accessed February 21, 2022, at https://www.epa.gov/sites/default/files/2015-08/documents/method_300-0_rev_2-1_1993.pdf).

US EPA Method 302.0 (2009): Determination of bromate in drinking water using two-dimensional IC with suppressed conductivity detection. United States Environmental Protection Agency. Office of Groundwater and Drinking Water, USEPA, Cincinnati. (Accessed February 21, 2022, at https://nepis.epa.gov).

US EPA Method 314.0 (1999) Determination of perchlorate in drinking water using IC. United States Environmental Protection Agency. Office of Groundwater and Drinking Water, USEPA, Cincinnati. (Accessed February 21, 2022, at https://nepis.epa.gov).

US EPA Method 314.1 (2005) Determination of perchlorate in drinking water using inline column concentration/matrix elimination IC with suppressed conductivity detection. United States Environmental Protection Agency. Office of Groundwater and Drinking Water, USEPA, Cincinnati. (Accessed February 21, 2022, at https://nepis.epa.gov).

US EPA Method 314.2 (2008) Determination of perchlorate in drinking water using two-dimensional IC with suppressed conductivity detection. United States Environmental Protection Agency. Office of Groundwater and Drinking Water, USEPA, Cincinnati. (Accessed February 21, 2022, at https://nepis.epa.gov).

US EPA Method 317.0 (2000) Determination of inorganic oxyhalide disinfection by-products in drinking water using ion chromatography with the addition of a postcolumn reagent for trace bromate analysis. United States Environmental Protection Agency. Office of Groundwater and Drinking Water, USEPA, Cincinnati. (Accessed February 21, 2022, at https://www.nemi.gov/methods/method_summary/4675).

US EPA Method 321.8 (1997) Determination of bromate in drinking waters by ion chromatography inductively coupled plasma – mass spectrometry. United States Environmental Protection Agency. Office of Groundwater and Drinking Water, USEPA, Cincinnati. (Accessed February 21, 2022, at https://www.epa.gov/sites/default/files/2015-09/documents/m_321_8.pdf).

US EPA Method 326.0 (2002) Determination of inorganic oxyhalide disinfection by-products in drinking water using IC incorporating the addition of a suppressor acidified postcolumn reagent for trace bromate analysis. United States Environmental Protection Agency. Office of Groundwater and Drinking Water, USEPA, Cincinnati. (Accessed February 21, 2022, at https://nepis.epa.gov).

US EPA Method 331.0 (2005) Determination of perchlorate in drinking water by liquid chromatography electrospray ionization mass spectrometry, Revision 1.0. United States Environmental Protection Agency. Office of Groundwater and Drinking Water, USEPA, Cincinnati. (Accessed February 21, 2022, at https://nepis.epa.gov).

US EPA Method 332 (2005) Determination of perchlorate in drinking water by IC with suppressed conductivity and electrospray ionization mass spectrometry. United States Environmental Protection Agency. Office of Groundwater and Drinking Water, USEPA, Cincinnati. (Accessed February 21, 2022, at https://nepis.epa.gov).

US EPA Method 557 (2009): Determination of haloacetic acids, bromate and dalapon in drinking water by IC electrospray ionization tandem mass spectrometry. United States Environmental Protection Agency. Office of Groundwater and Drinking Water, USEPA, Cincinnati. (Accessed February 21, 2022, at https://nepis.epa.gov).

US EPA Method 9056A (2007): Determination of inorganic anions by IC. United States Environmental Protection Agency. Office of Groundwater and Drinking Water, USEPA, Cincinnati. (Accessed February 21, 2022, at https://www.epa.gov/sites/default/files/2015-12/documents/9056a.pdf).

WHO (2020). Chromium in drinking-water, draft background document for development of WHO guidelines for drinking-water quality. World Health Organization, Geneva, Switzerland. (Accessed February 21, 2022, at https://apps.who.int/iris/bitstream/handle/10665/338062/WHO-HEP-ECH-WSH-2020.3-eng.pdf).

3.2 Beer

Beer is one of the oldest fermented alcoholic drinks, a very complex mixture of compounds which defines its sensorial properties. Beer contains ethanol, carbohydrates, proteins, magnesium, potassium, antioxidants, B vitamins, glycerol, organic acids etc. (De Keukeleire, 2000). A proper product quality is critical for consistent taste and other sensory characteristics that are essential for consumers' acceptance; hence the analytical methods used in quality control must be accurate and robust to fulfill many regulations. The analysis of organic acids, carbohydrates, sulfite, ethanol and inorganic ions are of primary interest (Bruce, 2002; Yan et al., 1997).

Among the broad range of techniques used for beer analysis (gas chromatography, high performance liquid chromatography, atomic absorption spectrometry, enzyme-based methods, wet chemical methods, etc.), in the last few years, IC has become the method of choice for several analytes. It is a valuable tool in the determination of:
- substances that are monitored to establish the extent of fermentation, as well as the quality of the end-product (carbohydrates, alcohols)
- compounds contributing to taste, such as inorganic ions and organic acids
- added preservatives and colorants (to ensure food safety)
- the quality of water used (inorganic ions)

Water is a primary ingredient used in brewing. IC monitoring of the water used in the brewing process for inorganic anions and cations is an important issue since they have a major impact on the taste of beer. This is discussed in detail in the previous subchapter.

The carbohydrates of most importance in the brewing industry are the fermentable ones, while oligosaccharides contribute to the caloric value of the beer. A proper knowledge of the carbohydrate profiles helps to control the brewing process and to improve the beer quality. This starts at the level of beer wort[11], in which the most important sugars for brewing are maltose and maltotriose – converted to ethanol by the added yeast. Carbohydrates can be determined using columns, such as Metrosep Carb 2, CarboPac PA1 or CarboPac PA 100, with pulsed amperometry for detection, taking between 30–60 min per separation (Table 11).

The determination of ethanol and glycerol are of noteworthy importance. These compounds have a considerable effect on the flavor and taste; they can be determined by ion-exclusion chromatography, such as in the case of using a Pac ICE-AS6 column, with 100 mM perchloric acid as the mobile phase and pulsed amperometry for detection (Thermo Scientific application note 46). When using an AminexR HPX 87H column and with refractive index detection, besides ethanol and glycerol, maltotriose, glucose, fructose and lactic acid can also be determined (Klein and Leubolt, 1993).

Organic acids from beer can be targeted to monitor fermentation, but they are also related to beer flavor. Among organic acids, oxalic acid can cause problems if it is present in high concentrations, since it can lead to the precipitation of calcium oxalate, reducing the quality of beer by making it cloudy. The analysis of these acids in beer can be achieved by anion-exchange chromatography using columns such as IonPac AS11, which enables the separations of both organic acids and inorganic anions (Figure 46), or by ion-exclusion chromatography on columns such as

---

11 Beer wort is the liquid in which malt starch is converted enzymatically into sugars and then is additionally flavored with hops.

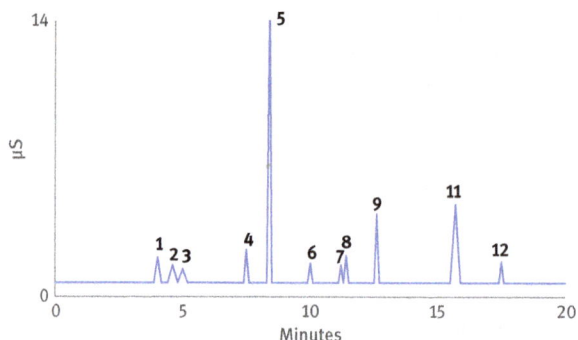

Figure 46: Separation of inorganic anions and organic acids on an IonPac AS11 column. Peak ID: 1, fluoride; 2, lactate; 3, acetate; 4, pyruvate; 5, chloride; 6, nitrate; 7, succinate; 8, malate; 9, sulfate; 10, phosphate; 11, citrate (Thermo Scientific application note 46).

IonPac ICE-AS6 with 0.4 mM heptafluorobutyric acid as the mobile phase, using suppressed conductivity detection (Table 11).

Most of the inorganic ions from beer originate from the water used during the brewing process and a strict quality control is required because the ionic composition influences the taste of the finished product. It was established that excessive amounts of certain ions can influence the fermentation and have adverse effects on the beer's taste (Briggs et al., 1981; Bruce, 2002):

- chloride enhances the beer's sweetness and may impede yeast flocculation
- sulfate is found naturally in water but can be added deliberately to bring out the hops' flavor ("burtonization")
- nitrate can cause problems if it is converted to nitrite, which can harm yeast metabolism, leading to weak or incomplete fermentation
- sodium can give a certain smoothness to the beer, when combined with chloride, but if too much sulfate is present, it gives an unpleasant harshness to the flavor
- potassium can give a salty taste, while potassium chloride leads to a bitter, astringent and soapy taste
- magnesium is an important nutrient for yeast, but magnesium sulfate can cause a bitter-sour taste at higher concentrations

A proper monitoring of the anion and cation profiles in beer is hence essential to ensure quality and meet the consumer needs in the brewing industry. Ion-exchange chromatography, with suppressed conductivity detection, is used for the determination for both anions and cations. Using an IonPac AS11 column and a gradient elution, besides inorganic anions (fluoride. chloride, nitrate, sulfate and phosphate), several organic acids can also be determined (lactate, acetate, pyruvate, succinate, malate and citrate), while an IonPac CS12 column can be successfully used for cations' analysis (Figure 47). Two channel systems enable the simultaneous determination of cations

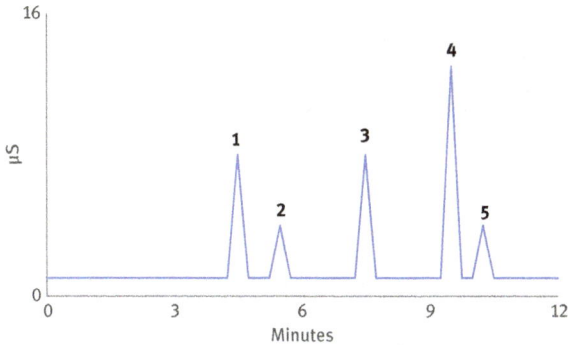

Figure 47: Separation of inorganic cations (1 – sodium, 2 – ammonium, 3 – potassium, 4 – magnesium, 5 – calcium) from lager on an IonPac CS12 column (Thermo Scientific application note 46).

and anions, such as in the case of the Metrohm's two-channel system (Metrohm application note D-002), in which cations are separated on a Metrosep C6 column and anions on a Metrosep A Supp 10 column (Figure 48).

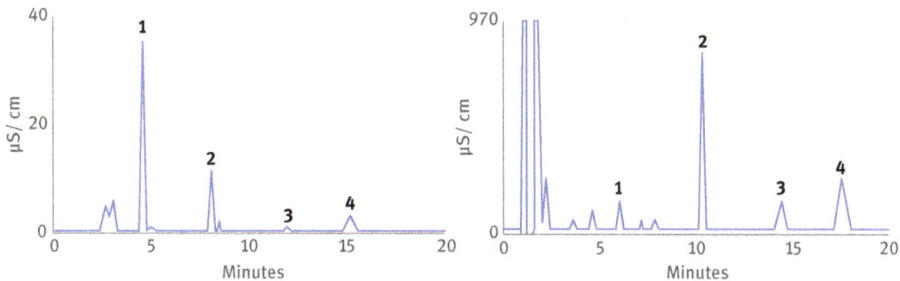

Figure 48: Simultaneous separation of inorganic anions (a: 1, chloride; 2, phosphate; 3, nitrite; 4, sulfate/Metrosep A Supp 10 column) and cations (b: 1, sodium; 1, 2, potassium; 3, magnesium; 4, calcium/Metrosep C6 column) from beer using a dual-channel instrument (Metrohm, IC application note D-002).

Among anions, sulfite has a distinct place. Sulfites are widely used as preservatives. They are added to foods and beverages to improve their shelf life by inhibiting the microbial growth, to enhance or preserve the color, and they also have antioxidative properties (Guido, 2016). Unfortunately, sulfite intake has been correlated to a wide range of adverse reactions (hypersensitivity, allergic reactions and vitamin deficiency leading to anaphylactic reactions), hence sulfites are now regulated and monitored in foods and beverages (Taylor et al., 1986). The Food and Agriculture Organization (FAO) and World Health Organization (WHO) Codex Committee on Food Labeling have listed sulfites among the food ingredients that must be declared as allergens (FAO/WHO, 2019). Ion-exclusion chromatography, followed by electrochemical detection, is the

Table 11: Summary of IC applications for beer analysis.

Analytes/matrix	Column	Mobile phase	Detection	Sample preparation	Separation time (min)	Reference
Glycerol and ethanol in beer	IonPac ICE-AS6	100 mM HClO$_4$ 1 mL/min	Pulsed amperometry	Degassing/ dilution/ filtration	12	Thermo Scientific application note 46
Ethanol, maltotriose, glucose, fructose, lactic acid, glycerol in nonalcoholic beer	Aminex HPX 42A	Water 0.5 mL/min	Refractive index	Degassing/ dilution/filtration	21	Klein and Leubolt (1993)
Fucose, fructose, saccharose and maltose in wort	Metrosep Carb 2	100 mM NaOH + 25 mM CH$_3$COONa/0.2 M NaOH + 0.22 M CH3COONa 0.7 mL/min	Pulsed amperometry	Degassing/ dilution/filtration	60	Metrohm, IC application note P-084
Fermentable sugars (glucose, fructose, isomaltose, sucrose, maltose, maltotriose) in wort	CarboPac PA1	Water/0.5 M NaOH 1 mL/min	Pulsed amperometry	Degassing/ dilution/filtration	30	Thermo Scientific application note 46
Malto-oligosaccharides (glucose, maltose, maltotriose, maltotetraose, maltopentaose, maltohexaoze, maltoheptaose, maltooctaose, maltodecaose) in beer	CarboPac PA 100	Water/0.5 M NaOH/1 M CH$_3$COONa 1 mL/min	Pulsed amperometry	Degassing/ dilution/filtration	30	Thermo Scientific application note 46

Analytes	Column	Mobile phase	Detection	Sample preparation		Reference
Mannitol, arabinose, glucose, fructose, lactose; sucrose, raffinose, maltose, acetate, glycolate, formate in beer	HPIC-AS6	80 mM NaOH 1 mL/min	Conductivity and pulsed amperometry	Dilution/filtration	12	Yan et al. (1997)
Organic acids (pyruvic, citric, malic, formic, lactic, acetic, succinic) in beer	IonPac ICE-AS6	0.8 mM heptafluorobutyric acid 0.8 mL/min	Pulsed amperometry	Degassing/dilution/filtration	24	Thermo Scientific application note 46
Anions and organic acids (fluoride. lactate, acetate, pyruvate, chloride, nitrate, succinate, malate, sulfate, phosphate, citrate) in ale	IonPac AS11	Water/1 mM+ NaOH/100 mM+ NaOH/ Methanol 2 mL/min	Suppressed conductivity	Degassing/dilution/filtration	20	Thermo Scientific application note 46
Anions (dihidrophosphate, chloride, bromide, nitrate, sulfate) and organic acids (acetic, succinic, pyroglutamic, lactic, pyruvic, oxalic, citric) in beer	Shodex IC I-I524 A	1.5mM phthalic acid + 1.38mM tris (hydroxymethyl) aminomethane + 300 mM boric acid 1.2 mL/min	Nonsuppressed conductivity	Degassing/dilution/filtration	20	Shodex application note – anions in beer
Organic acids (citric, pyruvic, gluconic, malic,succinic, lactic, fumaric, acetic, pyroglutamic) in beer	Shodex RSpak KC-G 6B + KC-811	4.8mM HClO$_4$ 1 mL/min	VIS (430 nm, post-column derivatization)	Degassing/dilution/filtration	30	Shodex application note KC 811
Oxalic acid + fluoride, chloride, nitrate, sulfate and phosphate in beer	Allsep Anion 100	0.85 mM NaHCO$_3$, 0.9 mM Na$_2$CO$_3$ 1.2 mL/min	Suppressed conductivity	Degassing/dilution/filtration	17	Marten (2003)

(continued)

Table 11 (continued)

Analytes/matrix	Column	Mobile phase	Detection	Sample preparation	Separation time (min)	Reference
Anions (chloride, nitrate, sulfate and phosphate) in beer	Metrohm A SUPP 5	Na_2CO_3/$NaHCO_3$	Nonsuppressed conductivity	Degassing/ dilution/filtration	24	Bruce (2002)
Anions (chloride, phosphate, nitrate and sulfate + sulfite) in beer	Metrosep A Supp 10	4 mM Na_2CO_3 + 6 mM $NaHCO_3$ + 5 µM $NaClO_4$ 0.7 mL/min	Suppressed conductivity	Degassing/ dilution/filtration	17	Metrohm, IC application note D-002
Sulfite +chloride, phosphate, sulfite, bromide, nitrate and sulfate in beer	Metrosep A Supp 10	6 mM Na_2CO_3 4 mM $NaHCO_3$ 5 µM $NaClO_4$ 0.7 mL/min	Suppressed conductivity	Dilution with a solution containing formaldehyde and NaOH/filtration	22	Metrohm, IC application note S-225
Sulfite	Metrosep Carb 2	0.3 M NaOH + 0.3M CH_3COONa 0.5 mL/min	Pulsed amperometry	Mixing with stabilization solution (1 M formaldehyde + 0.2 M NaOH)/ filtration	8	Espinosa (2020)

3.2 Beer — 105

Anions (fluoride, chloride, nitrate, phosphate, sulfate) in beer	Dionex Ion Pac AS 14 (250 × 4.6 mm))	3.5 mM Na_2CO_3+ 1 mM $NaHCO_3$ 1 mL/min	Suppressed conductivity	Degassing/dilution/filtration	–	Michalski et al. (2021)
Cations (potassium, sodium, magnesium, calcium) in beer	Metrosep C2-250 (250 × 4.6 mm),	8 mM tartaric acid + 1.5 mM 1,2-dipicolinic 0.9 mL/min	Nonsuppressed conductivity		–	
Cations (sodium, potassium, magnesium, calcium) in beer	Metrosep C6	2.3 mM HNO_3 + 1.7 mM dipicolinic acid 0.9 mL/min	Nonsuppressed conductivity	Dilution/filtration (ultrafiltration cell)	20	Metrohm IC application note D-002
Cations (potassium, sodium, magnesium, calcium) in lager	IonPac CS12	Water/100 mM methanesulfonic acid 1 mL/min	Suppressed conductivity	Degassing/ dilution/filtration	12	Thermo Scientific application note 46

method of choice of the Association of Official Analytical Chemists (AOAC Official Method 990.31, 1995) for the determination of this analyte, but it can also be separated on ion-exchange columns like Metrosep A Sup 10, which separates sulfite, besides other inorganic anions in 22 min, detection being accomplished using suppressed conductivity (Metrohm, IC application note S-225).

Table 11 summarizes several IC applications for beer analysis, emphasizing the most common detection methods in beer analysis:
– for cations' analysis: suppressed and nonsuppressed conductivity
– for anions' analysis: suppressed conductivity, nonsuppressed conductivity and pulsed amperometry
– for carbohydrates' analysis: pulsed amperometry

In most cases, the sample preparation required by beer analysis using IC is reduced to a minimum, consisting of degassing, filtration and dilution.

References

AOAC Official Method 990.31 (1995), Sulfites in foods and beverages, ion exclusion. (Accessed August 4, 2021, at http://www.aoacofficialmethod.org/index.php?main_page=product_info&cPath=1&products_id=2781.

Briggs, D. E., Hough, J. S., Stevens, R., & Young, T. W. (1981). Malting and brewing science: malt and sweet wort (Vol. 1). Springer Science & Business Media. 207, 225, 277.

Bruce, J. (2002). Analysis of anions in beer using ion chromatography. Journal of automated methods & management in chemistry, 24(4), 127–130.

De Keukeleire, D. (2000). Fundamentals of beer and hop chemistry. Quimica Nova, 23, 108–112.

Espinosa, A. L. M. (2020). A simplified method to determine total sulhite content in food and beverages via ion chromatography. The Column, 16(2),12–16.

FAO/WHO (2019) Codex Alimentarius: general Standard for Food Additives – Codex Standard 192–1995; Secretariat of the Joint FAO/WHO Food Standards Programme, FAO: Rome, Italy.

Guido, L. F. (2016). Sulfites in beer: reviewing regulation, analysis and role. Scientia Agricola, 73, 189–197.

Klein, H., & Leubolt, R. (1993). Ion-exchange high-performance liquid chromatography in the brewing industry. Journal of chromatography. A, 640(1–2), 259–270.

Marten, S. (2003). Ion chromatography determination of oxalic acid in beer. LC GC Europe, Suppl.S, 12–13.

Metrohm application note D-002, Anions and cations in beer; streamlining beverage analysis with ion chromatography. (Accessed August 4, 2021, at https://partners.metrohm.com/GetDocumentPublic?action=get_dms_document&docid=4864162).

Metrohm IC application note S-225, Sulfite besides standard anions in beer on the Metrosep A Supp 10. (Accessed August 4, 2021, at https://partners.metrohm.com/GetDocumentPublic?action=get_dms_document&docid=696036).

Metrohm, IC application note P-084, Determination of glucose, fructose, saccharose and maltose oligosaccharides in wort applying pulsed amperometric detection after dose-in gradient elution. (Accessed August 4, 2021, at https://partners.metrohm.com/GetDocumentPublic?action=get_dms_document&docid=3461202).

Michalski, R., Muntean, E., & Łyko, A. (2021). Major inorganic ions in Polish beers. Bulletin of University of Agricultural Sciences and Veterinary Medicine Cluj Napoca, Food Sciences and Technology, 78 (1), 48–56.

Shodex application note KC 811 – organic acids in beer. (Accessed May, 16, 2022, at https://www. shodex.com/en/dc/03/08/13.html).

Shodex application note – anions in beer. (Accessed May 16, 2022, at https://www.shodex.com/ en/dc/07/02/15.html).

Taylor, S. L., Higley, N. A., & Bush, R. K. (1986). Sulfites in foods: uses, analytical methods, residues, fate, exposure assessment, metabolism, toxicity and hypersensitivity. Advances in Food Research, 30, 1–76.

Thermo Scientific application note 46 (2016). Ion chromatography: a versatile technique for the analysis of beer. (Accessed August 4, 2021, at https://tools.thermofisher.com/content/sfs/bro chures/AN-46-IC-Beer-Analysis-AN71410-EN.pdf).

Yan, Z., Xingde, Z., & Weijun, N. (1997). Simultaneous determination of carbohydrates and organic acids in beer and wine by ion chromatography. Microchimica Acta, 127(3), 189–194.

3.3 Wine

Wine is an appreciated beverage, obtained by the alcoholic fermentation of white or red grapes. During fermentation, yeasts convert the fermentable sugars, mainly in ethanol and carbon dioxide; the resulting product is a complex matrix that contains a broad variety of organic and inorganic substances, such as ethanol, organic acids, aminoacids, carbohydrates, phenolic compounds, potassium, sodium, calcium and iron. Most of these are beneficial for human health; hence the recommendation for a moderate consumption of wines (especially the red ones) for improved health condition and longevity (Artero et al., 2015; Guilford and Pezzuto, 2011; Tsang et al., 2005), the "French paradox" being a well-known subject in thematic researches (Ferrières, 2004; Haseeb et al., 2017; Renaud and Lorgeril, 1992).

For obtaining a product with the desired properties and/or for quality control, chemical analyses are needed in wine production for a broad range of purposes, such as: to monitor the changes in concentrations of certain compounds during winemaking, to trace contaminations, to determine certain legal limits, to check the quality of the final product or even to confirm the authenticity. In this context, IC is used for the reliable, accurate and precise determination of numerous major and minor compounds that affect taste (e.g., carbohydrates, organic acids, anions, cations, biogenic amines), a synthetic image of the applications being presented in Table 12.

Carbohydrates are the major components in many wines. Of these, glucose and fructose are the main ones present in grapes; they are utilized by yeasts during fermentation to produce ethanol. According to the sugar content, wines are classified into dry, semi-dry, semi-sweet and sweet. Despite carbohydrates can be easily determined using high performance liquid chromatography, IC can also successfully separate these compounds as well as several organic acids, such as in the case of using a HPIC–AS column, with conductivity and pulsed amperometry detection, in less than twelve minutes, using minimal sample processing (Yan et. al., 1997). In fact, because all carbohydrates from wine are water-soluble, IC is particularly suitable for analyzing them; it is advantageous since it does not require extraction, hence the determination can be performed directly.

Organic acids are important components in enology. They play an important role in wine preservation and affect organoleptic properties. One can differentiate between acids originating from grape (such as tartaric, malic, citric) and those formed during alcoholic fermentation (e.g., succinic, lactic, acetic); the major ones are tartaric acid and malic acid, which bring the most important contribution to the acidity of wines. These compounds are valuable tracers for wine quality, taste and stability, helping to also track the alteration processes and the authenticity of wines . During wine maturation, certain changes in organic acid composition occur, such as in the case of free tartaric acid – its concentration decreases since it precipitates by reacting with other components of the wine. Malic acid can be converted to lactic acid by bacteria during malolactic fermentation to produce wine with a lower acidity and different taste. Organic acids also contribute to the overall acidity, tartness and flavor; malic acid can cause a green apple flavor whereas excessive acetic acid will cause an undesirable vinegar taste (Waterhouse et al., 2016). Higher concentrations of acids, such as acetic, propionic or lactic, occur in spoiled wines. All these acids can be determined by ion-exclusion chromatography and suppressed conductivity detection in numerous applications (Table 12), but because the peak capacity of certain ion-exclusion columns is low and some organic acids are not well resolved, in some instances, it is necessary to use columns in series. Unfortunately ion-exclusion does not allow the simultaneous determination of inorganic anions. On the contrary, ion-exchange separations allow the simultaneous determination of a large variety of organic acids, along with inorganic anions, in one run. Metrohm provides columns for two approaches: fast screening with a Metrosep A Supp 7 column, which provides a good separation for acetate, malate, tartrate and oxalate, besides major anions (chloride, phosphate, sulfite and sulfate) in less than 20 min or a more complex profiling by running a gradient on a Metrosep A Supp 7 column, which resolves gluconate, lactate, acetate, propionate, isobutyrate, butyrate, methacrylate, valerate, methyl sulfate, dichloroacetate, malonate, malate, glutarate, adipate and phthalate. One of the most impressive applications is one that was achieved on a Dionex IonPac AS11-HC-4 μm column, a high-resolution high-capacity anion-exchange product able to separate 22 organic acids besides eight inorganic anions (Chen et. Al, 2013).

Biogenic amines are another class of organic compounds of interest in wine. They are released during fermentative process, influencing the wine taste. Their concentrations in wines are variable and depend on the storage time and conditions, the quality of raw materials and on the microbial contamination during the wine-making process. Biogenic amines have been related to lack of hygiene or poor manufacturing practices (Guo et al., 2015). Besides, some of them can have undesirable effects on human health, such as nausea, respiratory distress, headache, sweating, heart palpitations and hyper- or hypotension (Feddern et al., 2019). The determination of these compounds is challenging due to their low concentrations and properties. By using gradient-suppressed cation chromatography, the separation of biogenic amines can be accomplished on Metrosep C Supp 1 column, together with several inorganic cations (Metrohm IC Application Note CS-014) or on IonPac CS18 column, using suppressed conductivity and/or pulsed amperometry. Suppressed conductivity allows the detection of most target biogenic amines, but pulsed amperometry provides a broader selectivity, enabling the detection of all biogenic amines of interest (De Borba and Rohrer, 2007; Metrohm IC Application Note CS-014).

Inorganic ions may originate from soil as well as from climatic conditions, pesticides, production and storage practices (Galani-Nikolakaki and Kallithrakas-Kontos, 2006; Maltman, 2008; White and White, 2003):

-   chloride content tends to be higher in sea regions, or where there is a tradition of washing barrels with salt water, but can be added to adjust wine taste
-   lithium is present in wine due to plant root uptake or due to storage in glass bottles. Due to its very low natural concentration, it is used in Italy as a denaturing agent to mark wines that are not allowed for consumption (Zerbinati et al., 2000)
-   potassium and calcium are natural components of grape – potassium concentration in wines is usually high and it can precipitate with tartaric acid, while the content of calcium can be affected by the addition of calcium carbonate for deacidification; hence a high level of this metal can lead to precipitation of calcium tartrate
-   sulfiting agents are added to prevent oxidation and bacterial contamination
-   certain inorganic ions may be toxic, such as nitrates, lead and cadmium
-   as the wine ages, the concentration of inorganic ions increases

Cations in wine can be separated by ion-exchange chromatography (e.g., using Shodex IC YS-50 or IonPac CS12 columns) with conductivity detection (Figure 49–50), with a minimal sample processing, which consists of dilution/filtration and, eventually, solid phase extraction (Shodex application note YS-50: Pyrzyńska, 2004; Zerbinati et al., 2000). The same mechanism also separates anions in columns, such as Metrosep A Supp 10, Metrosep Anion Dual 2, Shodex IC SI-90 4E (Metrohm IC Application Notes S–12, S–281, S-396; Shodex application note SI-90 4E); the most efficient applications are those in which both inorganic and organic anions are separated in the same run (Chen et al., 2013; Thermo Scientific/Dionex – application note 273). Again, minimal sample processing is necessary in most cases (dilution), but the use of in-line ultrafiltration is recommended since it protects the column from particulates and ensures a trouble-free operation (Metrohm IC Application Notes S–281, S-396).

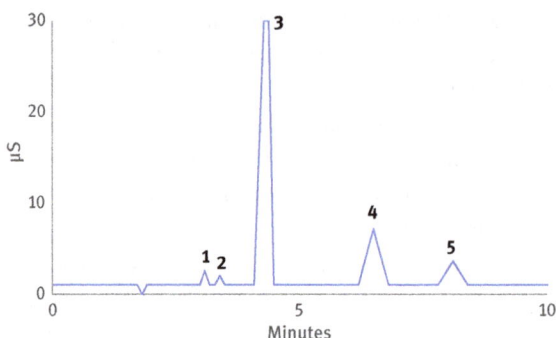

Figure 49: Separation of cations from red wine on an IonPac CS12A column. Peak ID: 1, sodium; 2, ammonium; 3, potassium; 4, magnesium; 5, calcium (Pyrzyńska, 2004).

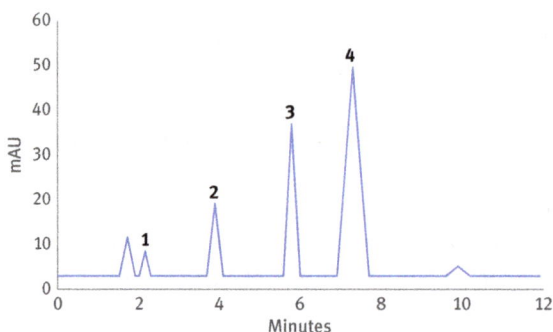

Figure 50: Separation of cations from red wine on a Metrosep C2 column using VIS detection after postcolumn derivatization. Peak ID: 1, copper; 2, zinc; 3, iron(II); 4, manganese (Metrohm IC Application Note C-105).

Sulfite has a distinct position among anions. The addition of sulfites is a practice widely used in winemaking because of their antioxidant and stabilizing effects. Sulfite is a widely used food preservative and antioxidant that has been generally recognized as safe (GRAS) for several decades. However, due to numerous reported adverse reactions, from 1986, the FDA has required warning labels on any food containing more than 10 mg/kg of sulfite or beverage containing more than 10 mg/L. Six sulfiting agents are currently approved by the FDA for use as food additives: sulfur dioxide, sodium sulfite, sodium and potassium bisulfite and sodium and potassium metabisulfite. Because sulfite is an unstable anion in solution and oxidizes to sulfate, samples and standards should be analyzed rapidly. Mannitol can be added to prevent such an oxidation (samples stable up to 24 h), while unpreserved samples should be analyzed as soon as possible. To avoid oxidation during the chromatographic separation, the ultrapure water used in analysis must be degassed prior to eluent preparation as well as for all other working

solutions and standards (Espinoza, 2020).The general analytical procedure requires an alkaline treatment of the sample to release the bound sulfite (because,
wine samples are acidic due to the presence of naturally occurring organic acids),
followed by either anion-exchange separation (Espinoza, 2020; Metrohm IC Application Notes S–12, S–281, S-396; Shodex application note SI-90 4E) or ion exclusion
chromatography (AOAC Official Method 990.31-1995; Chen et al., 2016; Kim, 1990).
A representative separation is depicted in Figure 51.

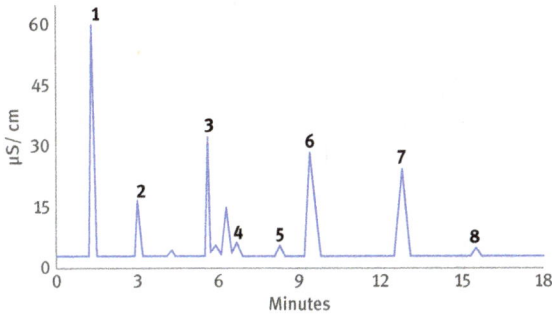

Figure 51: Separation of organic and inorganic anions from white wine on a Metrosep A Supp 10
column. Peak ID: 1, acetate; 2, chloride; 3, phosphate; 4, malate; 5, sulfite; 6, tartrate; 7, sulfate;
8, oxalate (Metrohm IC Application Notes S–281, S-396).

Using IC in wine analysis offers multiple advantages, such as: a wide linear range,
accurate results, no interferences, the possibility of analyzing organic and inorganic
anions at the same time, while the sample preparation procedure is simple and
quick.

Table 12: Summary of IC applications for wine analysis.

Analytes/matrix	Column	Mobile phase	Detection	Sample prep	Separation time (min)	Reference
Mannitol, arabinose, glucose, fructose, sucrose, raffinose, maltose, acetate, glycolate, formate in beer, wine	HPIC-AS6	80 mM NaOH 1 mL/min	Conductivity and pulsed amperometry	Dilution/ filtration	12	Yan et al. (1997)
Organic acids and anions in wine	Metrosep A Supp 16 (250 × 4 mm), Metrosep A Sup-7 (150 × 4 mm)	1 mM NaOH/60 mM NaOH – gradient 0.7 mL/min	Suppressed conductivity	Dilution	100	Metrohm IC Application Note S-362.
Organic acids (citric, tartaric, malic, succinic, lactic, fumaric, acetic) + carbonate in wine	2x HPICE-AS1 in series	2 mM octanesulfonic acid in 2% 2-propanol 0.5 mL/min	Suppressed conductivity	Filtration/solid phase extraction/ dilution	60	Thermo Scientific/ Dionex – application note 21
Organic acids (citric, tartaric, malic, succinic and lactic, acetic) + carbonate in wine– fast run	HPICE-AS1	2 mM octanesulfonic acid in 2% 2-propanol 0.8 mL/min	Suppressed conductivity	Filtration/solid phase extraction/ dilution	15	Thermo Scientific/ Dionex – application note 21

Analytes	Column	Eluent	Detection	Sample preparation		Reference
Organic acids (9) + 7 inorganic anions in fruit juices and wines	OmniPac PAX-100	Water/12% methanol/16% ethanol/0.1–1 M NaOH – gradient 1 mL/min	Suppressed conductivity		35	Thermo Scientific/ Dionex – application note 273.
Organic acids (22) + 8 inorganic anions in fruit juices, wines	Dionex IonPac AS11-HC-4 µm, Dionex IonPac AS11-HC-9 µm	KOH (Dionex EGC 500 Eluent Generator Cartridge with Dionex CR-ATC 500 continuously regenerated anion trap column) 0.4 mL/min	Suppressed conductivity	Dilution/ Dionex OngGuard II RP cartridge	45	Chen et al. (2013)
Gluconate, lactate, acetate, propionate, isobutyrate, butyrate, methacrylate, valerate, methyl sulfate, dichloroacetate, malonate, malate, glutarate, adipate and phthalate in wine	Metrosep A Supp 7	Ultrapure water/6.4 mM Na_2CO_3 + 2 mM $NaHCO_3$ 0.7 mL/min	Suppressed conductivity	–	90	Metrohm IC Application Note S-396
Acetate, chloride, phosphate, malate, **sulfite,** tartrate, sulfate and oxalate in wine (fast screening)	Metrosep A Supp 10	5 mM Na_2CO_3 + 5 mM $NaHCO_3$ + 5 µM $HClO_4$ 1 mL/min	Suppressed conductivity	Inline ultrafiltration	18	Metrohm IC Application Notes S–281, S-396
Lactate, chloride, nitrate, **sulfite** and phosphate in wine	Metrosep Anion Dual 2	2 mM $NaHCO_3$, 1.8 mM Na_2CO_3, 15% acetone 0.8 mL/min	Suppressed conductivity	Dilution with 0.2 mM NaOH containing HCHO for sulfite's stabilization	18	Metrohm IC Application Note S–12

(continued)

Table 12 (continued)

Analytes/matrix	Column	Mobile phase	Detection	Sample prep	Separation time (min)	Reference
Cations (sodium, potassium, calcium, magnesium) and biogenic amines (putrescine, cadaverine and histamine) in wine	Metrosep C 1	2.5 mM HNO3, 10% acetone	Direct conductivity	Dilution with 2 mM HNO_3	28	Metrohm IC Application Note C-70
Copper, zinc, iron(II) and manganese in wine	Metrosep C 2	1.75 mM oxalic acid + 2 mM ascorbic acid 1 mL/min	UV/VIS (after post-column reaction with 0.2 mM PAR, 2.4 M ammonia, 1 M acetic acid)	Dilution with 20 mM ascorbic acid	12	Metrohm IC Application Note C-105.
Biogenic amines (dopamine, tyramine, putresceine, cadaverine, histamine, serotonin agmantine, phenylethylamine, spermidine, spermine) in wine	IonPac CS18	3–45 mM methanesulfonic acid/ gradient 0.3 mL/ min	Pulsed amperometry, suppressed conductivity	Dilution/ centrifugation/ filtration	42	De Borba and Rohrer (2007); Thermo Scientific/ Dionex – application note 182

Analyte/sample	Column	Eluent/flow	Detection	Sample preparation		Reference
Biogenic amines (monomethylamine, triethylamine, 2-phenylethylamine, putresceine, cadaverine, histamine, serotonin) besides cations (sodium, potassium, calcium, magnesium)/red wine	Metrosep C Supp 1	Gradient with 2.5 mM HNO$_3$ + 100 µg/L rubidium/25 mM HNO$_3$ + 100 µg/L rubidium 1 mL/min	Suppressed conductivity	Dilution	30	Metrohm IC Application Note CS-014.
Lithium (+sodium, potassium, magnesium and calcium)/wine	IonPac CS12	20 mM methanesulfonic acid 1 mL/min	Suppressed conductivity	SPE on C18 cartridges	15	Zerbinati et al. (2000)
Cations (sodium, ammonium, potassium, magnesium and calcium/wine	IonPac CS12A	20 mM methanesulfonic acid 1 mL/min	Conductivity	–	10	Pyrzyńska (2004)
Cations (sodium, ammonium, potassium, magnesium, calcium) and monoethanolamine/red wine	Shodex IC YS-50	4 mM methanesulfonic acid 1 mL/min	Nonsuppressed conductivity	Dilution	10	Shodex application note YS-50
Sulfite in wine	IonPac ICE-AS1	20 mM methanesulfonic acid 0.2 mL/min	Pulsed amperometry	Dilution with buffer containing mannitol/ filtration	20	Chen et al. (2016)
Sulfite + anions (chloride, nitrate, hydrogenphosphate, sulfate)/wine	Shodex IC SI-90 4E	1mM Na2HCO3 + 4 mM NaHCO$_3$ + 5% acetone 1.5 mL/min	Suppressed conductivity	Dilution	18	Shodex application note SI-90 4E

(continued)

Table 12 (continued)

Analytes/matrix	Column	Mobile phase	Detection	Sample prep	Separation time (min)	Reference
Sulfite in wine	Shodex RS pack KC-811	12 mM H_3PO_4 1.2 mL/min	Electrochemical detection	Dilution	16	Shodex application note – KC-811
Sulfite in wine	Metrosep Carb 2	0.3 M NaOH + 0.3 M CH_3COONa 0.5 mL/min	Pulsed amperometry	Mixing with stabilization solution (1 M HCHO + 0.2 M NaOH)/filtration	8	Espinosa (2020)

References

AOAC Official Method 990.31-1995, Sulfites in foods and beverages. Ion exclusion. in Official Methods of Analysis of AOAC International, 16th ed., Vol. II; Cunniff, P., Ed.

Artero, A., Artero, A., Tarín, J. J., & Cano, A. (2015). The impact of moderate wine consumption on health. Maturitas, 80(1), 3–13.

Chen, L., De Borba, B., & Rohrer, J. (2013). Determination of organic acids in fruit juices and wines by high-pressure IC. Thermo Fisher Scientific application note 1068. (Accessed August 11, 2021, at https://assets.thermofisher.cn/TFS-Assets/CMD/Application-Notes/AN-1068-IC-Organic-Acids-Fruit-Juice-Wine-AN70753-EN.pdf).

Chen, L., De Borba, B., & Rohrer, J. (2016). Determination of total and free sulfite in foods and beverages. Thermo Fisher Scientific application note 54. (Accessed August 11, 2021, at http://www.thermoscientific.es/content/dam/tfs/ATG/CMD/cmd-documents/sci-res/app/chrom/ic/col/AN-54-IEX-Sulfite-Food-Beverage-AN70379-EN.pdf).

De Borba, B. M., & Rohrer, J. S. (2007). Determination of biogenic amines in alcoholic beverages by ion chromatography with suppressed conductivity detection and integrated pulsed amperometric detection. Journal of chromatography. A, 1155(1), 22–30.

Espinosa, A. L. M. (2020). A simplified method to determine total sulphite content in food and beverages via ion chromatography. The Column, 16(2),12–16.

Feddern, V., Mazzuco, H., Fonseca, F. N., & De Lima, G. J. M. M. (2019). A review on biogenic amines in food and feed: toxicological aspects, impact on health and control measures. Animal Production Science, 59(4), 608–618.

Ferrières, J. (2004). The French paradox: lessons for other countries. Heart, 90(1), 107–111.

Galani-Nikolakaki, S. M., & Kallithrakas-Kontos, N. G. (2006). Elemental content of wines. Mineral Components in Foods, 323–344.

Guilford, J. M., & Pezzuto, J. M. (2011). Wine and health: a review. American Journal of Enology and Viticulture, 62(4), 471–486.

Guo, Y. Y., Yang, Y. P., Peng, Q., & Han, Y. (2015). Biogenic amines in wine: a review. International Journal of Food Science & Technology, 50(7), 1523–1532.

Haseeb, S., Alexander, B., & Baranchuk, A. (2017). Wine and cardiovascular health: a comprehensive review. Circulation, 136(15), 1434–1448.

Kim, H. J. (1990). Determination of sulfite in foods and beverages by ion exclusion chromatography with electrochemical detection: collaborative study. Journal of the Association of Official Analytical Chemists, 73(2), 216–222.

Maltman, A. (2008). The role of vineyard geology in wine typicity. Journal of Wine Research, 19(1), 1–17.

Metrohm IC Application Note C-70. Cations and biogenic amines in wine. (Accessed August 11, 2021, at https://partners.metrohm.com/GetDocumentPublic?action=get_dms_document&docid=694833).

Metrohm IC Application Note C-105. Copper, zinc, iron(II) and manganese in wine by ion chromatography with post-column reaction and UV/VIS detection. (Accessed August 11, 2021, at https://partners.metrohm.com/GetDocumentPublic?action=get_dms_document&docid=694650).

Metrohm IC Application Note CS-014. Biogenic amines besides other cations in red wine applying a high-pressure gradient. (Accessed August 11, 2021, at https://partners.metrohm.com/GetDocumentPublic?action=get_dms_document&docid=3087146).

Metrohm IC Application Note S–12. Determination of lactate, chloride, nitrate, sulfite and phosphate in wine. (Accessed August 11, 2021, at https://partners.metrohm.com/GetDocumentPublic?action=get_dms_document&docid=696334).

Metrohm IC Application Note S–281 (2009). Inorganic and organic anions in wine applying inline ultrafiltration. (Accessed May 17, 2022, at https://www.metrohm.com/content/dam/metrohm/shared/application-files/AN-S-281.pdf).

Metrohm IC Application Note S–362. Organic acid anions in wine applying a low-pressure gradient. (Accessed May 17, 2022, at https://www.metrohm.com/content/dam/metrohm/shared/application-files/AN-S-362.pdf).

Metrohm IC Application Note S-396 (2021). Assessing wine quality with IC. Organic acid analysis using suppressed conductivity detection. (Accessed August 11, 2021, at https://partners.metrohm.com/GetDocumentPublic?action=get_dms_document&docid=4869637).

Pyrzyńska, K. (2004). Analytical methods for the determination of trace metals in wine. Critical Reviews in Analytical Chemistry, 34(2), 69–83.

Renaud, S. D., & de Lorgeril, M. (1992). Wine, alcohol, platelets and the French paradox for coronary heart disease. The Lancet, 339(8808), 1523–1526.

Shodex application note – KC-811. Sulphite in wine. (Accessed August 11, 2021, at https://www.shodex.com/en/dc/07/07/03.html).

Shodex application note SI-90 4E. Sulphite in wine. (Accessed August 11, 2021, at https://www.shodex.com/en/dc/07/05/14.html#!).

Shodex application note YS-50. Cations in red wine. (Accessed August 11, 2021, at https://www.shodex.com/en/dc/07/03/44.html).

Thermo Scientific/Dionex – application note 21. Organic acids in wine. (Accessed August 11, 2021, at https://assets.thermofisher.com/TFS-Assets/CMD/Application-Notes/4118-AN21_LPN032025-02.pdf).

Thermo Scientific/Dionex – application note 182. Determination of biogenic amines in alcoholic beverages by ion chromatography with suppressed conductivity and integrated pulsed amperometric detections. (Accessed August 11, 2021, at http://www.cromlab.es/Articulos/Columnas/HPLC/Thermo/Dionex/CS18/56196-AN182_IC_BiogenicAmines_Alcohol_14Aug2007_LPN1888_02.pdf).

Thermo Scientific/Dionex – application note 273. Higher resolution separation of organic acids and common inorganic anions in wine. (Accessed August 11, 2021, at https://assets.thermofisher.com/TFS-Assets/CMD/Application-Notes/AN-273-IC-Organic-Acids-Inorganic-Anions-Wine-LPN2727-EN.pdf).

Tsang, C., Higgins, S., Duthie, G. G., Duthie, S. J., Howie, M., Mullen, W., & Crozier, A. (2005). The influence of moderate red wine consumption on antioxidant status and indices of oxidative stress associated with CHD in healthy volunteers. British Journal of Nutrition, 93(2), 233–240.

Waterhouse, A. L., Sacks, G. L., & Jeffery, D. W. (2016). Understanding wine chemistry. John Wiley & Sons.

White, R. E., & White, R. E. (2003). Soils for fine wines. Oxford University Press.

Yan, Z., Xingde, Z., & Weijun, N. (1997). Simultaneous determination of carbohydrates and organic acids in beer and wine by ion chromatography. Microchimica Acta, 127(3), 189–194.

Zerbinati, O., Balduzzi, F., & Dell'Oro, V. (2000). Determination of lithium in wines by ion chromatography. Journal of chromatography. A, 881(1–2), 645–650.

3.4 Milk and dairy products

Milk is one of the foods containing most of the basic nutrients in optimal proportions and with high bioavailability. It is an indispensable food in the initial months of life because it contains all the elements that the body needs in the early stages after birth. Later, its great quality is the relatively high calcium content in a form that is compatible with the requirements of the human body. Milk and dairy products are recommended for a wide group of consumers, including children, adolescents and the elderly, respecting the condition of not being consumed in excess. The consumption of whole milk is excluded or limited only in a relatively small number of cases, such as in acute enterocolitis or in the aggravation of chronic ones, accompanied by diarrhea, in the pre- and post-operative period, in congenital intolerance or acquired after diseases of the gastrointestinal tract due to insufficient lactase in the intestine. Dairy products obtained after lactic acid fermentation or after the introduction of specific microorganisms in milk are important in the diet, because acidity increases in the fermented products due to the lactic acid synthesized from lactose, the proteins partially disintegrate, while the amount of B vitamins increases. Compared to raw milk, fermented milk products assimilate more easily, stimulate the secretion of the digestive glands, normalize intestinal peristalsis and inhibit putrefactive and harmful microbes (Akram et al., 2020; Nagpal et al., 2012; Tsompo, 2018).

Carbohydrates are important macrocomponents of milk. Lactose is the major sugar in natural milk, but other carbohydrates, such as galactose, glucose, fructose, sucrose and maltose, are also found in dairy products. A proper quantification of mono- and disaccharide is important for food processing units to ensure product formulation and product quality and to report ingredients to immune and allergy-sensitive individuals. They can be determined by IC using Dionex CarboPac columns, detection being accomplished with pulsed amperometry (Cataldi et al., 2003; Christison et al., 2014; Hu and Rohrer, 2020) as shown in Table 13 and Figure 52.

High-performance anion-exchange chromatography, coupled with pulsed amperometric detection (HPAE-PAD), is one of the most powerful techniques for carbohydrate determination, combining the high selectivity and efficiency of chromatographic separation with the high detection sensitivity and specificity of amperometric detection (Brunt et al., 2021; Monti et al., 2017), being mentioned also as an ISO method (ISO 22184, 2021). HPAE-PAD is a direct detection technique, hence eliminating the possible errors associated with analyte derivatization.

Determination of residual lactose is an important issue in the quality control of products that are declared lactose-free, since a number of lactase-deficient individuals have difficulty in digesting the lactose in milk products (a health condition known as "lactose intolerance"), which can lead to bloating, abdominal pain and diarrhea after drinking milk. The increasing demand for lactose-free products needs for an accurate, reliable and sensitive method to analyze them – such as those using Metrosep Carb 2 or CarboPac PA20 columns, the separation being accomplished under alkaline conditions, followed by pulsed amperometric detection (Perati et al., 2016). Because milk is a difficult matrix that is rich in protein, it must undergo dialysis before analysis. A more demanding separation can be accomplished with a Dionex CarboPac PA210 column using a Dionex EGC 500 KOH eluent generator cartridge and dual detection: pulsed amperometry and mass spectrometry (Aggrawal and Rohrer, 2018).

Anions of interest in dairy products are iodide, thiocyanate and perchlorate; these need to be determined for health and hygiene reasons. Small amounts of iodine are necessary for the normal development of human beings, but excess can

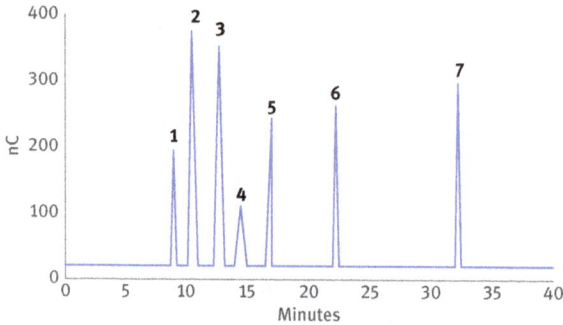

Figure 52: Separation carbohydrates from milk on a CarboPac PA1 column. Peak ID: 1, arabinose; 2, galactose; 3, glucose; 4, sucrose; 5, fructose; 6, lactose; 7, maltose (Hu and Rohrer, 2020).

lead to thyroid disorders. Important amounts of iodine are present in table salt, seafood, as well as in dairy products. Milk and dairy products are among the most common food sources of iodine (Murray et al, 2008; Pearce, 2007). Besides, iodide-containing compounds are used as disinfectants in the dairy industry. IC, coupled with pulsed amperometric detection, can be used to determine iodide in milk products, using an IonPac AS11 column, with nitric acid as the mobile phase, in less than five minutes. Samples have to be diluted, treated with CH3COOH 3%, filtrated and then passed through OnGuard-RP cartridges to remove fat (Table 13).

Mineral salts in milk are mainly found in the form of chlorides, phosphates and nitrates (salts of magnesium, calcium, iron, copper, manganese). These salts are of special importance to the growing organism, participating in the formation and strengthening of bones. They are involved in the process of coagulating the milk and in ensuring the obtaining of a suitable curd for the manufacture of cheese (Foroutan et al., 2019; Manuelian et al., 2018; Zamberlin et al., 2012). IC plays a crucial role in analyzing anions and cations from milk and dairy products, two representative chromatograms being represented in Figures 53 and 54 (Cataldi et al., 2003; Gaucheron and Le Graet, 2000; Metrohm application note S–179).

Choline is a valuable micronutrient from milk that plays a vital role in the proper functioning of the liver, brain, nervous system and in metabolism. It is involved in maintaining cardiovascular and liver health, in reproduction and development and in also helping in memory and physical performance. Rich sources of choline are milk, eggs and organ meats. Because it is essential for human metabolism and is critical for the function and synthesis of the neurotransmitter acetylcholine, choline is often added to vitamin formulations, animal feeds, infant formulas and sports drinks as bitartrate or chloride salt (Hollenbeck, 2012; Wallace et al., 2018; Ziesel, 2004). IC is an important alternative for choline determination from milk, simultaneously with other cations of alkali and alkaline-earth metals, using an IonPac CS12A column and suppressed conductivity detection. The samples have to be digested with HCl at 70° C to release the bounded choline from milk, then filtrated and diluted (Dionex/Thermo Scientific, Application Note 124).

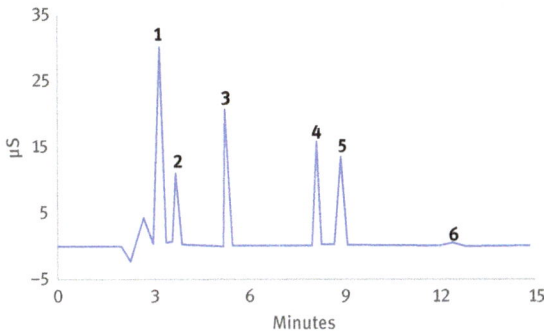

Figure 53: Separation of anions from hard cheese on an IonPac AS11-HC column, with 20 mM NaOH at 0.38 mL/min. Peak ID: 1, chloride; 2, sulfate; 3, bromide (internal standard); 4, phosphate; 5, citrate; 6, isocitrate (Cataldi et al., 2003).

Milk whey is the liquid that remains after the manufacture of cheese, after the separation of the curd. It is a byproduct and also a valuable raw material as its composition offers numerous commercial uses (Macwan et al, 2016; Mollea et al., 2013; Spalatelu, 2012). The quality control of whey and its derivatives is a very demanding field, in which IC proves to be useful for the determination of carbohydrates, anions and cations (Cataldi et al., 2003). The obtained data indicates that whey contains large amounts of important ions at higher concentration levels, in comparison to milk. The content of ions in whey have a strong impact on the design of technological processes for its increased utilization.

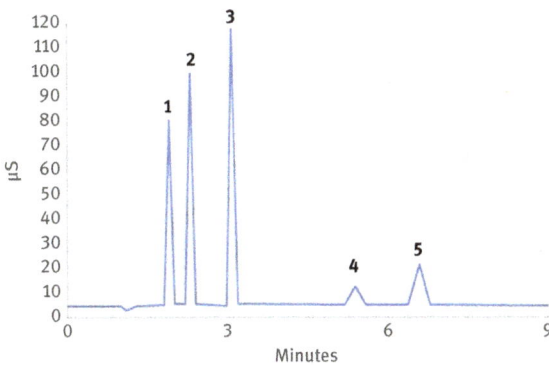

Figure 54: Separation of cations from hard cheese on a IonPac CS12A column, with 20 mM methanesulfonic acid at 1 mL/min. Peak ID: 1, lithium (internal standard); 2, sodium; 3, potassium; 4, magnesium; 5, calcium (Cataldi et al., 2003).

Table 13: Summary of IC applications in the analysis of milk and dairy products.

Analytes/matrix	Column	Mobile phase	Detection	Sample prep	Separation time (min)	Reference
Carbohydrates (sucrose, galactose, glucose, lactose and lactulose) in milk and dairy products	CarboPac PA20	10 mM KOH 0.008 mL/min	Pulsed amperometry	Dilution/deproteinization with Carrez reagents/dilution/centrifugation/filtration/OnGuard IIA cartridge	12	Christison et al. (2014)
Carbohydrates (arabinose, galactose, glucose, sucrose, fructose, lactose and maltose) in dairy products	CarboPac PA1	1 M CH_3COONa / 0.2 M $NaOH/H_2O$/ 25 mM CH_3COONa (gradient) 0.25 mL/min	Pulsed amperometry	Mixing with methanol+ethanol +buffer/deproteinization with Carrez reagents/dilution/centrifugation/filtration	35	Hu and Rohrer (2020)
Carbohydrates (galactose, glucose, N-acetylgalactosamine, lactose, lactulose and epilactose) in milk whey	CarboPac PA1	10 mM NaOH + 2 mM $Ba(OAc)_2$ 1 mL/min	Pulsed amperometry	Deproteinization with Carrez reagents/centrifugation/dilution/filtration	20	Cataldi et al. (2003)
Lactose, glucose, fructose, lactic acid in lactic acid beverages and fermented milk	Shodex sugar SH-G+ SH1011	5 mM H_2SO_4	Refractive index, UV (210 nm)	Dilution/deproteinization with sulfosalicylic acid/centrifugation/filtration	10	Shodex application note – saccharides in lactic acid beverages and fermented milk
Residual lactose in lactose-free milk	Metrosep Carb 2 -150/4.0	5 mM NaOH/2 mM CH_3COONa 0.8 mL/min	Pulsed amperometry	Dilution, in-line dialysis	20	Metrohm, IC application note P-55

Analyte	Column	Eluent/flow	Detection	Sample preparation		Reference
Residual lactose+ arabinose, fucose galactose, glucose, fructose, sucrose, alolactose, lactulose, epilactose and raffinose in lactose-free dairy products	Dionex CarboPac PA210	KOH/eluent generator 0.8 mL/min	Pulsed amperometry and mass spectrometry	Dilution/deproteinization with Carrez reagents/ centrifugation/OnGuard IIA cartridge	12	Aggrawal and Rohrer (2018)
Residual lactose + galactose, glucose, sucrose, lactulose in lactose-free milk products	CarboPac PA20	$0.2\ M$ NaOH, $0.1\ M$ CH_3COONa, $1\ M$ CH_3COONa 0.5 mL/min	Pulsed amperometry	Dilution/deproteinization with Carrez reagents/ centrifugation/OnGuard IIA cartridge	30	Perati et al. (2016)
Residual lactose + galactose, fructose, sucrose and lactulose in lactose-free milk and dairy products	CarboPac PA20-Fast -4µm	$0.2\ M$ NaOH, $0.1M$ CH_3COONa, $1\ M$ CH3COONa 0.25 mL/min	Pulsed amperometry	Dilution/deproteinization with Carrez reagents/ filtration	15	Thielel and Jensen (2018)
Iodide in milk products	IonPac AS11 Analytical, 4 × 250	$50\ mM$ HNO_3 1.5 mL/min	Pulsed amperometry	Dilution/protein precipitation with CH_3COOH 3%/filtration/ SPE	4	Dionex/Thermo Scientific, Application Note 37
Iodide in milk powder	Metrosep A Supp 5	$1\ mM$ $NaHCO_3$ $3.2\ mM$ Na_2CO_3 0.7 mL/min	Suppressed conductivity	Extraction in water/in-line dialysis	30	Metrohm IC application note S–162

(continued)

Table 13 (continued)

Analytes/matrix	Column	Mobile phase	Detection	Sample prep	Separation time (min)	Reference
Choline in dry milk and infant formula	IonPac CG12A	18 mM H_2SO_4 1 mL/min	Suppressed conductivity	Treatment with HCl at 70 °C/ centrifugation/filtration/ dilution	25	Dionex/ Thermo Scientific, Application Note 124
Choline (+sodium, ammonium, potassium, calcium, magnesium) in baby milk powder	Metrosep C Supp 1	4 mM HNO_3 50 µg/L rubidium 1 mL/min	Suppressed conductivity	Treatment with HCl at 70 °C/ centrifugation/filtration/pH adjustment/dilution	28	Metrohm, application note CS-004
Iodide, thiocyanate and perchlorate in milk	Metrosep A Supp 15-50/4.0	4 mM Na_2CO_3/6 mM $NaHCO_3$, 10% MeOH 0.8 mL/min	Suppressed conductivity	In-line dialysis	28	Metrohm, IC Application Note S–297
Anions (chloride, sulfate, bromide, phosphate, citrate and isocitrate) in milk whey	IonPac AS11-HC	20 mM NaOH 0.38 mL/min	Suppressed conductivity	Deproteinization with Carrez reagents/centrifugation/ dilution/filtration	15	Cataldi et al. (2003)
Anions (hydrogen carbonate, chloride, nitrate, hydrogen phosphate) in milk	Shodex IC I-524A	1.5 mM p-hydroxybenzoic acid + 1.7 mM N,N-diethylethanolamine + 10% CH_3OH 1.5 mL/min	Nonsuppressed conductivity	Dilution/SPE	20	Shodex application note – anions in milk

Analyte/matrix	Column	Eluent/flow	Detection	Sample preparation	(min)	Reference
Cations (lithium, sodium, ammonium, potassium, calcium, magnesium) in milk whey	IonPac CS12A	20 mM methanesulfonic acid 1 mL/min	Suppressed conductivity	Centrifugation/dilution/filtration	10	Cataldi et al. (2003)
Cations and lactic acid in whey powder	Metrosep C 6	4 mM HNO_3 0.9 mL/min	Direct conductivity	Extraction in eluent/heating at 50 °C/filtration	35	Metrohm, IC application note S–179
Ammonium (+ lithium, sodium, ammonium, potassium, calcium, magnesium) in milk and dairy products	IonPac CS15	5 mM H_2SO_4 + 9% CH_3CN 1.2 mL/min	Suppressed conductivity	Extraction with water and HCl/filtration. Centrifugation/filtration/dilution – for yoghurt	40	Gaucheron and Le Graet (2000)

References

Aggrawal, M., & Rohrer, J. (2018). Determination of lactose in lactose-free dairy products using HPAE coupled with PAD and MS dual detection. Thermo Scientific, Application Note 72780. (Accessed July 17, 2021, at https://assets.thermofisher.com/TFS-Assets/CMD/Application-Notes/an-72780-ic-ms-lactose-free-dairy-an72780-en.pdf).

Akram, M., Sami, M., Ahmed, O., Onyekere, P. F., & Egbuna, C. (2020). Health benefits of milk and milk products. In Functional Foods and Nutraceuticals (pp. 211–217). Springer.

Brunt, K., Sanders, P., Ernste-Nota, V., & Van Soest, J. (2021). Results of multi-laboratory trial ISO/CD 22184 – IDF/WD 244: milk and milk products – determination of the sugar contents – high-performance anion exchange chromatography method with pulsed amperometric detection. Journal of AOAC International, 104(3), 732–756.

Cataldi, T. R., Angelotti, M., D'Erchia, L., Altieri, G., & Di Renzo, G. C. (2003). Ion-exchange chromatographic analysis of soluble cations, anions and sugars in milk whey. European Food Research and Technology, 216(1), 75–82.

Christison, T., Verma, M., Kettle, A., Fisher, C., & Lopez, L. (2014). Determination of carbohydrates in beverages and milk products by HPAE-PAD and capillary HPAE-PAD. Thermo Scientific Poster Note PN71091_HPLC_2014_E_05/14S Thermo Fisher Scientific, Sunnyvale, CA, USA. (Accessed July 17, 2021, at https://tools.thermofisher.cn/content/sfs/posters/PN-71091-Carbohydrates-Beverages-Milk-HPAE-PAD-HPLC-2014-PN71091-EN.pdf).

Dionex/Thermo Scientific, Application Note 124, 2002, Determination of choline in dry milk and infant formula. (Accessed July 17, 2021, at https://assets.thermofisher.com/TFS-Assets/CMD/Application-Notes/4208-AN124_LPN1054-01.pdf).

Dionex/Thermo Scientific, Application Note 37, 2004, Determination of iodide in milk products. (Accessed July 17, 2021, at http://www.cromlab.es/Articulos/Columnas/HPLC/Thermo/Dionex/AS11/4128-AN37_24Jul95_LPN0702-03.pdf).

Foroutan, A., Guo, A. C., Vazquez-Fresno, R., Lipfert, M., Zhang, L., Zheng, J., & Wishart, D. S. (2019). Chemical composition of commercial cow's milk. Journal of Agricultural and Food Chemistry, 67(17), 4897–4914.

Gaucheron, F., & Le Graet, Y. (2000). Determination of ammonium in milk and dairy products by ion chromatography. Journal of chromatography. A, 893(1), 133–142.

Hollenbeck, C. (2012). An introduction to the nutrition and metabolism of choline. Central Nervous System Agents in Medicinal Chemistry, 12(2),100–113.

Hu, J., & Rohrer, J. (2020). Determination of sugars in dairy products using HPAE-PAD. Thermo Scientific, Application Note 73341. (Accessed July 17, 2021, at https://assets.thermofisher.com/TFS-Assets/CMD/Application-Notes/an-73341-ic-hpae-pad-sugars-dairy-products-an73341-en.pdf).

ISO 22184:2021, Milk and milk products-determination of the sugar contents – High performance anion exchange chromatography with pulsed amperometric detection method. (Accessed April 20, 2022, https://www.iso.org/standard/72822.html).

Macwan, S. R., Dabhi, B. K., Parmar, S. C., & Aparnathi, K. D. (2016). Whey and its utilization. International Journal of Current Microbiology and Applied Sciences, 5(8), 134–155.

Manuelian, C. L., Penasa, M., Visentin, G., Zidi, A., Cassandro, M., & De Marchi, M. (2018). Mineral composition of cow milk from multibreed herds. Animal Science Journal, 89(11), 1622–1627.

Metrohm, IC application note CS-004, Determination of choline in baby milk powder. (Accessed July 17, 2021, at https://partners.metrohm.com/GetDocumentPublic?action=get_dms_document&docid=2259008).

Metrohm, IC application note P-55, Residual lactose in lactose-free milk. (Accessed July 17, 2021, at https://partners.metrohm.com/GetDocumentPublic?action=get_dms_document&docid=1975953).

Metrohm, IC application note S–162, Iodide in milk powder. (Accessed July 17, 2021, at https://partners.metrohm.com/GetDocumentPublic?action=get_dms_document&docid=696481).

Metrohm, IC application note S–179, Cations and lactic acid in whey powder applying two separation mechanisms in the same analysis. (Accessed July 17, 2021, at https://partners.metrohm.com/GetDocumentPublic?action=get_dms_document&docid=%20%202994121).

Metrohm, IC application note S–297, Iodide, thiocyanate and perchlorate in milk applying inline dialysis. (Accessed July 17, 2021, at https://partners.metrohm.com/GetDocumentPublic?action=get_dms_document&docid=983897).

Mollea, C., Marmo, L., & Bosco, F. (2013). Valorization of cheese whey, a by-product from the dairy industry. In Food industry. IntechOpen (Accessed April 20, 2022, https://cdn.intechopen.com/pdfs/42000/InTech-Valorisation_of_cheese_whey_a_by_product_from_the_dairy_industry.pdf).

Monti, L., Negri, S., Meucci, A., Stroppa, A., Galli, A., & Contarini, G. (2017). Lactose, galactose and glucose determination in naturally "lactose-free" hard cheese: HPAEC-PAD method validation. Food Chemistry, 220, 18–24.

Murray, C. W., Egan, S. K., Kim, H., Beru, N., & Bolger, P. M. (2008). US Food and Drug Administration's total diet study: dietary perchlorate and iodine. Journal of Exposure Science & Environmental Epidemiology, 18, 571–580.

Nagpal, R., Behare, P. V., Kumar, M., Mohania, D., Yadav, M., Jain, S., & Yadav, H. (2012). Milk, milk products and disease free health: an updated overview. Critical Reviews in Food Science and Nutrition, 52(4), 321–333.

Pearce, E. (2007). National trends in iodine nutrition: Are we getting enough? Thyroid, 17, 823–827.

Perati, P., De Borba, B., & Rohrer, J. (2016). Determination of lactose in lactose-free milk products by high-performance anion-exchange chromatography with pulsed amperometric detection. Thermo Fisher Scientific. Application Note 248. (Accessed July 17, 2021, at https://assets.thermofisher.com/TFS-Assets/CMD/Application-Notes/AN-248-IC-Lactose-Milk-AN70236-EN.pdf).

Shodex application note – anions in milk. (Accessed July 17, 2021, at https://www.shodex.com/en/dc/07/02/40.html).

Shodex application note – saccharides in lactic acid beverages and fermented milk. (Accessed July 17, 2021, at https://www.shodex.com/en/dc/03/02/32.html).

Spalatelu, C. (2012). Biotechnological valorization of whey. Innovative Romanian Food Biotechnology, 10, 1–15.

Thielel, K. J. F., & Jensen, D. (2018). Fast determination of lactose in dairy products. Thermo Fisher Scientific. Application Note 72633. (Accessed July 17, 2021, at https://assets.thermofisher.com/TFS-Assets/CMD/Application-Notes/can-72633-ic-lactose-dairy-products-can72633-en.pdf).

Tsopmo, A. (2018). Phytochemicals in human milk and their potential antioxidative protection. Antioxidants, 7(2), 32–40.

Wallace, T. C., Blusztajn, J. K., Caudill, M. A., Klatt, K. C., Natker, E., Zeisel, S. H., & Zelman, K. M. (2018). Choline: the underconsumed and underappreciated essential nutrient. Nutrition Today, 53(6), 240.

Zamberlin, Š., Antunac, N., Havranek, J., & Samaržija, D. (2012). Mineral elements in milk and dairy products. Mljekarstvo, 62(2), 111–125.

Zeisel, S. H. (2004). Nutritional importance of choline for brain development. Journal of the American College of Nutrition, 23(sup6), 621S–626S.

3.5 Ion chromatographic applications in food authentication

Food authenticity is a major issue for consumers, regulators, producers and processors, because it may negatively impact the entire food chain. Fraudulent practices can negatively affect consumer confidence and safety, as well as businesses. Common food frauds include adulteration, expiry date extension, mislabeling using unapproved processes, false declaration of geographical/species and botanical or varietal origin of raw materials (Munekata et al., 2021; Robson et al., 2021). Food adulteration is an illegal practice that is motivated by economic gain, involving the addition of certain substances to food products or the substitution of some components with cheaper ones. The high demand for certain food products, for which the supply cannot meet the demand is another cause for this practice – e.g., there is more demand for Scotch whiskey sold in the world than is produced in Scotland, hence they are lots of mislabeled liquors coming from different other parts of the world (Aylott and McKenzie, 2010; Stupak et al., 2018). From food safety perspective, adulteration lowers the nutritional value of fruit juices, while certain ingredients may cause allergies, major health problems or may be even fatal, such as is in the case of melamine in infant formulas (Poonia et al., 2017; Yang et al., 2009). The leading food products with reported cases of food frauds include olive oil, honey, maple syrup, fruit juices, coffee, spices, milk and milk-based products (Everstine et al., 2013; Moore et al. 2012; Tibola et al., 2018).

In the last few decades, food authenticity has become a well-established field of research, which involves a complimentary approach using multiple analytical techniques (especially chromatography and spectroscopy), because a large amount of data is common in such a context. Chemometric analysis has become a widely used approach (Cozzolino, 2016; Guyon et al., 2013; He et al., 2021; Jha et al., 2016; Ogrinc et al., 2003; Olivery et al., 2016; Schieber, 2018).

For the case of fruit juices, the main authenticity issues are those caused from using cheaper and lower quality ingredients in their preparation instead of those declared on their labels. As the main components of fruit juices are water and the carbohydrates (fructose, glucose and sucrose) (Eisele and Drake 2005), juices' adulteration is usually accomplished by diluting with water, by using cheaper ingredients (mainly different combinations of sugar solutions and syrups), by adding food additives to mask an inferior quality (e.g., food colorants to enhance the color), by blending cheaper with more expensive ones, by using another type of fruit (or partially replacing one type of fruit juice/pulp with one that is less expensive, e.g., by the addition of apple juice to expensive berry juices/pulps or fruit preparations) or by the addition of peel and/or pulp wash (Dasenaki, 2010; Muntean, 2010). The most frequently used method for fruit juice adulteration is diluting a concentrated juice with water, adding a sweetener and then adjusting the flavor and color, as necessary. Carbohydrates' addition is compulsory to maintain the correct Brix value, the major carbohydrate profile, and the expected taste. Inexpensive commercially available sweeteners, such as inverted beet or cane syrup, high fructose corn syrup, hydrolyzed inulin syrup or sucrose, are common in such cases (Kelly and Downey 2005).

Fruit juices' authenticity can be established by fingerprinting methods, targeting several different types of substances (e.g., polyphenolic compounds, organic acids, carotenoids, aminoacids, anthocyanins, sugars, caffeine, theophylline and theobromine), but adulteration detection often requires the use of many complimentary analytical methods, such UV–VIS spectroscopy, flame or graphite furnace atomic absorption spectroscopy, capillary electrophoresis, gas chromatography, mass spectrometry or nuclear magnetic resonance spectroscopy (Chen et al. 1998; Fidelis et al. 2017; Lachenmeyer et al., 2005; Muntean 2010; Navarro-Pasqual-Ahuir et al., 2015; Ogrinc et al. 2003; Versari et al. 1997). Among these, liquid chromatography still has a strong position in the quantification of carbohydrates that are present in the added sweeteners, but not in the fruit of origin, as well as in establishing the presence of other specific compounds. Measuring carbohydrates in fruit juices is important not only for establishing the authenticity but also for assessing the quality and for checking the possible microbiological alteration during storage. besides, it also has a significant effect on the sensory properties and nutritional values of fruit products and has to be considered carefully for diabetic patients (Karadeniz and Eksi 2002).

IC is a convenient tool in establishing fruit juices' authenticity. Since a lot of adulterations are accomplished by adding different sweeteners, such as partially inverted sucrose, corn syrups or sucrose, oligosaccharides' profiles can be used to detect adulteration of natural fruit juices by comparing the chromatographic profiles of carbohydrates of genuine juices with suspicious ones. Oligosaccharide fingerprinting can also be helpful in detecting adulteration by examining the samples for the presence of sugars that are not normally detected in juices but are found in sugar syrups, or by calculating the fructose/glucose ratio and/or the glucose: fructose: sucrose distribution and comparing these with reference values. In the particular case of apple juices, a genuine product should contain a minimum fructose/glucose ratio of 1.6 and a maximum sucrose content of 3.5% (Karadeniz and Eksi 2002). For this approach, several chromatographic procedures (some of them involving ion-exchange separations) with refractive index detection or pulsed amperometric detection are available (Cataldi et al., 2000, Martinez Montero et al., 2004, Zielinski et al., 2014).

Anions – especially organic ones – can be valuable fingerprints to assess whether blending of expensive juices was accomplished using cheaper ones (Chinnici et al., 2005, Mato et al., 2005). The profile of organic acids being distinct for each type of fruit juice, the evidence of such a practice can be obtained by comparing the IC fingerprint of an authentic juice with that of a suspicious sample (Saccani et al., 1995). Cation composition of fruit juices is a less studied topic, being usually used as a quality control and/or research tool (Frankowski 2014, Lanzerstorfer et al., 2014, Muntean 2010, Swallow et al., 1991, White and Cancalon, 1992; Will and Dietrich 2013).

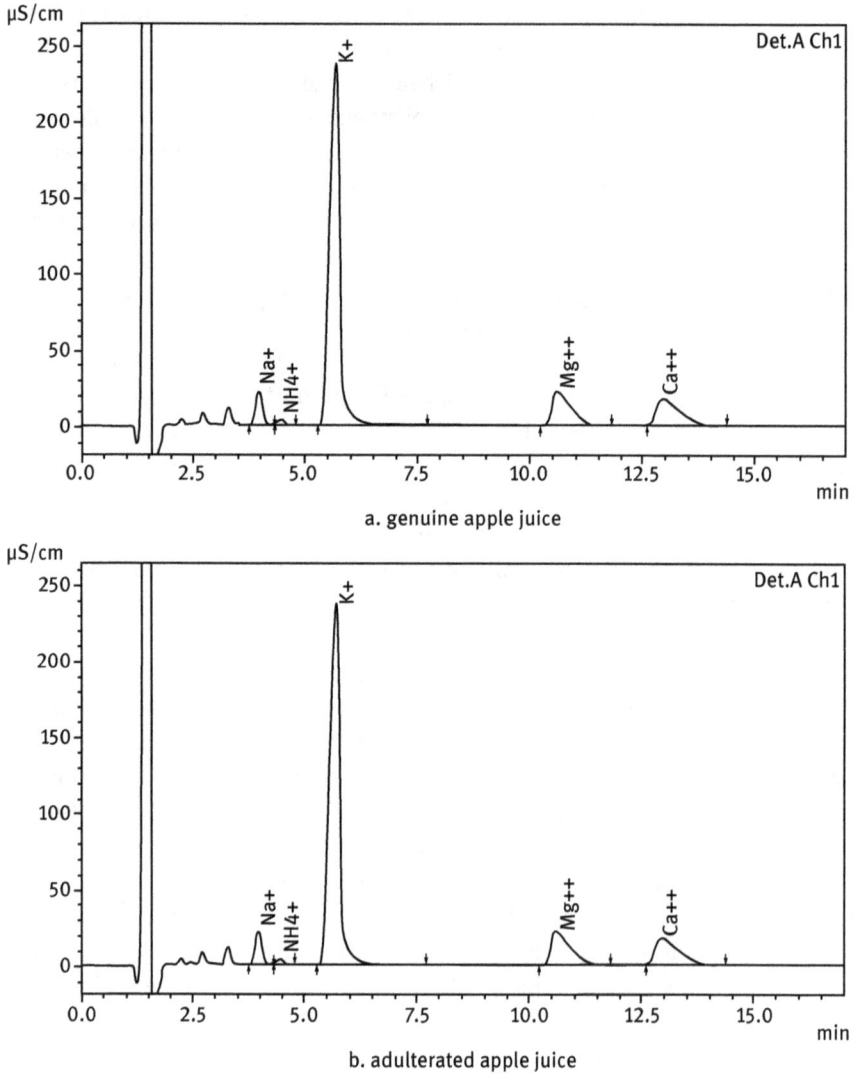

Figure 55: Comparative IC fingerprints for apple juices, recorded on a Shimadzu system, using a Universal Cation 7u column (100 × 4.6 mm–Alltech Associates), with 0.5 mL/min 3 mM HNO_3 – nonsuppressed conductivity detection.

This issue is significant but more difficult to use as a decision tool for authentication, since during fruit juices' preparation, some inorganic cations are introduced with water; others are introduced as counter ions to the added ingredients, while the fruit matrix contains its own cations (Figure 55).

Authentication of alcoholic beverages is another field in which IC has been reported to have made significant contribution, besides gas chromatography, HPLC,

Table 14: Summary of IC applications in food authentication.

Analytes/matrix	Column	Mobile phase	Detection	Sample prep	Separation (time (min)	Reference
Anions (nitrate, sulfate, oxalate) in Mexican agave spirits	Metrosep A Supp 5, (250 × 4 mm)	3.2 mM Na_2CO_3 + 1 mM $NaHCO_3$, 0.7 mL/min	Suppressed conductivity	Dilution/in-line dialysis	–	Lachenmeier et al. (2006)
Anions (nitrate, sulfate, oxalate) in tequila	Metrosep A Supp 5, (100 × 4 mm)	3.2 mM Na_2CO_3 + 1 mM $NaHCO_3$, 0.7 mL/min	Suppressed conductivity	Dilution/in-line dialysis	16	Lachenmeier et al. (2005)
Anions (chloride, nitrate and sulfate) in rum and vodka	IonPac AS4A, (250 × 4 mm)	17 mM Na_2CO_3 + 18 mM $NaHCO_3$, 2 mL/min	Suppressed conductivity	Evaporation/dilution	7	Lachenmeier et al. (2003)
Anions (chloride, nitrate and sulfate) in vodka	Akvilain A1 (250 × 4.6 mm)	1.8 mM Na_2CO_3 + 1.7 mM $NaHCO_3$, 1.2 mL/min	Conductivity		30	Arbuzov et al. (2002)
Cations (potassium, sodium, ammonium and lithium) in vodka	Akvilain C1 (250 × 4.6 mm)	4 mM HNO_3 2 mL/min	Conductivity		10	Arbuzov et al. (2002)
Carbohydrates (mannitol, sorbitol, glucose, fructose, sucrose) in juices	CarbopacPA-1 (250 × 4 mm)	0.15 NaOH/0.15 NaOH + 0.4 mM CH_3COONa (gradient)	Electrochemical detection	Dilution/filtration	8	An et al. (2021)

(continued)

Table 14 (continued)

Analytes/matrix	Column	Mobile phase	Detection	Sample prep	Separation (time (min)	Reference
Carbohydrates (glucose, fructose, sucrose) in fruit juices	2 × Carbo Pac PA1 (250 × 4 mm)	0.1 M NaOH/3 mM CH$_3$COONa (gradient) 0.7 mL/min	Pulsed amperometry	Ion exchange	80	Swallow et al (1991)
Cations (sodium, potassium, ammonium, magnesium and calcium) in apple and orange juices	Universal Cation 7u column (100 × 4.6 mm)	3 mM HNO$_3$ 0.5 mL/min	Nonsuppressed conductivity	Dilution/ filtration	15	Muntean (2010)
Glycerol, sucrose, glucose, fructose and organic acids (gluconic, lactic, malic, tartaric, citric) in beverages	IonPAC AS15 (25 × 2 mm)	10–90 mM KOH (gradient) 0.3 mL/min	Suppressed conductivity, mass spectrometry	Dilution/ filtration	40	Guyon et al. (2013)
Organic acids (oxalic, citric, galacturonic, dehydroascorbic, malic, quinic, ascorbic, succinic acid, shikimic acid and fumaric)and carbohydrates (sucrose, glucose and fructose) in fruit juices	Aminex HPX 87H (300×7.8 mm)	0.005 N H$_3$PO$_4$, 0.4 mL/min	UV and Refractive index detection	Ion exchange	20	Chinnici et al. (2005)

| Organic acids and anions in fruit juices | OmniPac Pax-500 (250 × 4 mm) | 60 mM NaOH in water–ethanol–methanol (66.5:20:13.5, v/v)/20 mM NaOH in water-ethanol (65:35, v/v)/60 mM NaOH in water-ethanol (65:35, v/v)/ gradient, 1 mL/min) | Suppressed conductivity | Dilution/filtration | 35 | Saccani et al. (1995) |

mass spectrometry, near-infrared spectrometry, isotope analysis, electronic nose and so on (Lachenmeier et al., 2003; Lachenmeyer, 2008). Illicit spirits originate mainly by blending high quality distillates with ethanol, which is made from a cheaper raw material, adding synthetic volatile components to alcohol or by mislabeling the variety and origin of the raw material (Lachenmeyer et al., 2008). Table 14 provides several directions in which IC can be a useful tool in food authentication.

References

An, J. A., Lee, J., Park, J., Auh, J. H., & Lee, C. (2021). Authentication of pomegranate juice using multidimensional analysis of its metabolites. Food Science and Biotechnology, 30(13), 1635–1643.

Arbuzov, V. N., & Savchuk, S. A. (2002). Identification of vodkas by ion chromatography and gas chromatography. Journal of Analytical Chemistry, 57(5), 428–433.

Aylott, R. I., & MacKenzie, W. M. (2010). Analytical strategies to confirm the generic authenticity of Scotch whisky. Journal of the Institute of Brewing, 116(3), 215–229.

Cataldi, T. R., Campa, C., & De Benedetto, G. E. (2000). Carbohydrate analysis by high-performance anion-exchange chromatography with pulsed amperometric detection: the potential is still growing. Fresenius' Journal of Analytical Chemistry, 368(8), 739–758.

Chen, Q. C., Mou, S. F., Hou, X. P., & Ni, Z. M. (1998). Simultaneous determination of caffeine, theobromine and theophylline in foods and pharmaceutical preparations by using ion chromatography. Analytica Chimica Acta, 371(2–3), 287–296.

Chinnici, F., Spinabelli, U., Riponi, C., & Amati, A. (2005). Optimization of the determination of organic acids and sugars in fruit juices by ion-exclusion liquid chromatography. Journal of Food Composition and Analysis, 18(2–3), 121–130.

Cozzolino, D. (2016). Near infrared spectroscopy and food authenticity. In Advances in Food Traceability Techniques and Technologies Woodhead Publishing, 119–136.

Dasenaki, M. E., & Thomaidis, N. S. (2019). Quality and authenticity control of fruit juices-a review. Molecules, 24(6), 1014.

Eisele, T. A., & Drake, S. R. (2005). The partial compositional characteristics of apple juice from 175 apple varieties. Journal of Food Composition and Analysis, 18(2–3), 213–221.

Everstine, K., Spink, J., & Kennedy, S. (2013). Economically motivated adulteration (EMA) of food: common characteristics of EMA incidents. Journal of Food Protection, 76(4), 723–735.

Fidelis, M., Santos, J. S., Coelho, A. L. K., Rodionova, O. Y., Pomerantsev, A., & Granato, D. (2017). Authentication of juices from antioxidant and chemical perspectives: A feasibility quality control study using chemometrics. Food Control, 73, 796–805

Frankowski, M. (2014). Aluminium and its complexes in teas and fruity brew samples, speciation and ions determination by ion chromatography and high-performance liquid chromatography–fluorescence. Food Analytical Methods, 7(5), 1109–1117.

Guyon, F., Gaillard, L., Brault, A., Gaultier, N., Salagoïty, M. H., & Médina, B. (2013). Potential of ion chromatography coupled to isotope ratio mass spectrometry via a liquid interface for beverages authentication. Journal of chromatography. A, 1322, 62–68.

He, Y., Bai, X., Xiao, Q., Liu, F., Zhou, L., & Zhang, C. (2021). Detection of adulteration in food based on nondestructive analysis techniques: A review. Critical Reviews in Food Science and Nutrition, 61(14), 2351–2371.

Jha, S. N., Jaiswal, P., Grewal, M. K., Gupta, M., & Bhardwaj, R. (2016). Detection of adulterants and contaminants in liquid foods – a review. Critical Reviews in Food Science and Nutrition, 56 (10), 1662–1684.

Karadeniz, F., & Ekşi, A. (2002). Sugar composition of apple juices. European Food Research and Technology, 215(2), 145–148.

Kelly, J. D., & Downey, G. (2005). Detection of sugar adulterants in apple juice using Fourier transform infrared spectroscopy and chemometrics. Journal of Agricultural and Food Chemistry, 53(9), 3281–3286.

Lachenmeier, D. W., Schmidt, B., & Bretschneider, T. (2008). Rapid and mobile brand authentication of vodka using conductivity measurement. Microchimica Acta, 160(1), 283–289.

Lachenmeier, D. W., Attig, R., Frank, W., & Athanasakis, C. (2003). The use of ion chromatography to detect adulteration of vodka and rum. European Food Research and Technology, 218(1), 105–110.

Lachenmeier, D. W., Richling, E., López, M. G., Frank, W., & Schreier, P. (2005). Multivariate analysis of FTIR and ion chromatographic data for the quality control of tequila. Journal of Agricultural and Food Chemistry, 53(6), 2151–2157.

Lachenmeier, D. W., Sohnius, E. M., Attig, R., & López, M. G. (2006). Quantification of selected volatile constituents and anions in Mexican Agave spirits (Tequila, Mezcal, Sotol, Bacanora). Journal of Agricultural and Food Chemistry, 54(11), 3911–3915.

Lanzerstorfer, P., Wruss, J., Huemer, S., Steininger, A., Müller, U., Himmelsbach, M., & Weghuber, J. (2014). Bioanalytical characterization of apple juice from 88 grafted and nongrafted apple varieties grown in upper Austria. Journal of Agricultural and Food Chemistry, 62(5), 1047–1056.

Martínez Montero, C., Dodero, R., Guillén Sánchez, D. A., & Barroso, C. G. (2004). Analysis of low molecular weight carbohydrates in food and beverages: a review. Chromatographia, 59(1), 15–30.

Mato, I., Suárez-Luque, S., & Huidobro, J. F. (2005). A review of the analytical methods to determine organic acids in grape juices and wines. Food Research International, 38(10), 1175–1188.

Moore, J. C., Spink, J., & Lipp, M. (2012). Development and application of a database of food ingredient fraud and economically motivated adulteration from 1980 to 2010. Journal of Food Science, 77(4), R118–R126.

Munekata, P. E., Domínguez, R., Pateiro, M., & Lorenzo, J. M. (2021). Introduction to food fraud. In Food Toxicology and Forensics, Academic Press, 1–30.

Muntean, E. (2010). Simultaneous carbohydrate chromatography and unsuppressed ion chromatography in detecting fruit juices adulteration. Chromatographia, 71(1), 69–74.

Navarro-Pascual-Ahuir, M., Lerma-García, M. J., Simó-Alfonso, E. F., & Herrero-Martínez, J. M. (2015). Quality control of fruit juices by using organic acids determined by capillary zone electrophoresis with poly (vinyl alcohol)-coated bubble cell capillaries. Food Chemistry, 188, 596–603.

Ogrinc, N. K. I. J., Košir, I. J., Spangenberg, J. E., & Kidrič, J. (2003). The application of NMR and MS methods for detection of adulteration of wine, fruit juices and olive oil. A review. Analytical and Bioanalytical Chemistry, 376(4), 424–430.

Oliveri, P., & Simonetti, R. (2016). Chemometrics for food authenticity applications. In Advances in food authenticity testing. Woodhead Publishing, 701–728.

Poonia, A., Jha, A., Sharma, R., Singh, H. B., Rai, A. K., & Sharma, N. (2017). Detection of adulteration in milk: A review. International Journal of Dairy Technology, 70(1), 23–42.

Robson, K., Dean, M., Haughey, S., & Elliott, C. (2021). A comprehensive review of food fraud terminologies and food fraud mitigation guides. Food Control, 120, 107516.

Saccani, G., Gherardi, S., Trifirò, A., Bordini, C. S., Calza, M., & Freddi, C. (1995). Use of ion chromatography for the measurement of organic acids in fruit juices. Journal of chromatography. A, 706(1–2), 395–403.

Schieber, A. (2018). Introduction to food authentication. In Modern techniques for food authentication. Academic Press, 1–21.

Stupak, M., Goodall, I., Tomaniova, M., Pulkrabova, J., & Hajslova, J. (2018). A novel approach to assess the quality and authenticity of Scotch Whisky based on gas chromatography coupled to high resolution mass spectrometry. Analytica Chimica Acta, 1042, 60–70.

Swallow, K. W., Low, N. H., & Petrus, D. R. (1991). Detection of orange juice adulteration with beet medium invert sugar using anion-exchange liquid chromatography with pulsed amperometric detection. Journal of the Association of Official Analytical Chemists, 74(2), 341–345.

Tibola, C. S., da Silva, S. A., Dossa, A. A., & Patrício, D. I. (2018). Economically motivated food fraud and adulteration in Brazil: incidents and alternatives to minimize occurrence. Journal of Food Science, 83(8), 2028–2038.

Versari, A., Biesenbruch, S., Barbanti, D. A., & Farnell, P. J. (1997). Adulteration of fruit juices: dihydrochalcones as quality markers for apple juice identification. LWT-Food Science and Technology, 30(6), 585–589.

White, R. D. Jr, & Cancalon, P. E. (1992). Detection of beet sugar adulteration of orange juice by liquid chromatography/pulsed amperometric detection with column switching. Journal of AOAC International, 75(3), 584–587.

Will, F., & Dietrich, H. (2013). Processing and chemical composition of rhubarb (*Rheum rhabarbarum*) juice. LWT-Food Science and Technology, 50(2), 673–678.

Yang, R., Huang, W., Zhang, L., Thomas, M., & Pei, X. (2009). Milk adulteration with melamine in China: crisis and response. Quality Assurance and Safety of Crops & Foods, 1(2), 111–116.

Zielinski, A. A. F., Braga, C. M., Demiate, I. M., Beltrame, F. L., Nogueira, A., & Wosiacki, G. (2014). Development and optimization of a HPLC-RI method for the determination of major sugars in apple juice and evaluation of the effect of the ripening stage. Food Science and Technology, 34, 38–43.

4 Recent developments in ion chromatography

Food science and food industry benefit from IC applications, which are used from the quality control and characterization of raw materials and ingredients, monitoring technological processes streams, up to the quality control of end-products, as well as in food safety or in establishing the authenticity of certain food products. Targeting for higher speed, sensitivity, efficiency and resolution, for bypassing sample limitations, recent trends in IC have been directed toward designing new stationary phases with smaller particles, packed in columns with lower internal diameters (nano and capillary columns), as well as in hyphenated techniques (e.g., the extremely expensive IC-MS, which therefore is not widely available), developing new sample preparation techniques or even new system design. From these, reagent-free IC, capillary IC and combustion IC will be presented.

4.1 Reagent-free ion chromatography

Mobile phases are usually prepared manually, this procedure leading in some cases to inconsistent concentrations and/or the contamination with certain ions (e.g., carbonate contamination of a hydroxide eluent) and finally to variability in retention times. Fortunately, the development of electrolytic eluent generation (EG) improved this situation, since by using this approach, mobile phases can be prepared in-line within the so-called reagent-free ion chromatography (RFIC) (Haddad et al., 2008; Liu et al., 2007; Srinivasan, 2021), this being one of the newest direction in IC. RFIC systems provide high performance with increased sensitivity and the flexibility to perform isocratic and gradient separations using alkali hydroxides, methanesulfonic acid or carbonate/bicarbonate eluents.

EG is based on electrolysis and can produce in-line high-purity eluent used for IC (Strong et al., 1991); using a combination of electrolysis of water with a membrane-based transport, the eluent ions are transferred across a membrane from a reservoir containing a high concentration of ions into a stream of deionized water in response to an applied current, hence the eluent is generated in-line, the accurate concentration of the mobile phase being determined by the ratio of current and the flow rate:

$$\text{Concentration} \sim \frac{\text{Current}}{\text{Flow rate}}$$

EG cartridges (EGC) work on the principle "just add water"; by applying a certain potential between the EGC's electrodes, the desired composition of the eluent is obtained. EGC convert electrolytically ultrapure water in KOH solutions of the required concentration: the ultrapure water is pumped through an electrolysis chamber where the following transformation occurs:

https://doi.org/10.1515/9783110644401-004

$$2H_2O + 2e^- \rightarrow 2OH^- + H_2$$

the generated hydroxide ions being then associated with potassium cations provided by the EGC. EGC can produce KOH, carbonate/bicarbonate or methanesulfonic acid according to the desired mobile phase composition. Isocratic or gradient
separations are possible since an EG can be programmed to produce either a constant composition (for isocratic separations) or a variable one (by driving water
through it and varying the current, one can achieve gradient elution); in both
cases, a high reproducibility can be obtained. In fact, the development of EG was
the key step which led to the widespread incorporation of gradient elution into IC,
with a corresponding increase in peak capacity. The composition of eluents resulted
from EG has greater accuracy, greater precision, being easier to prepare than eluents made by traditional bench techniques (Haddad et al., 2008).

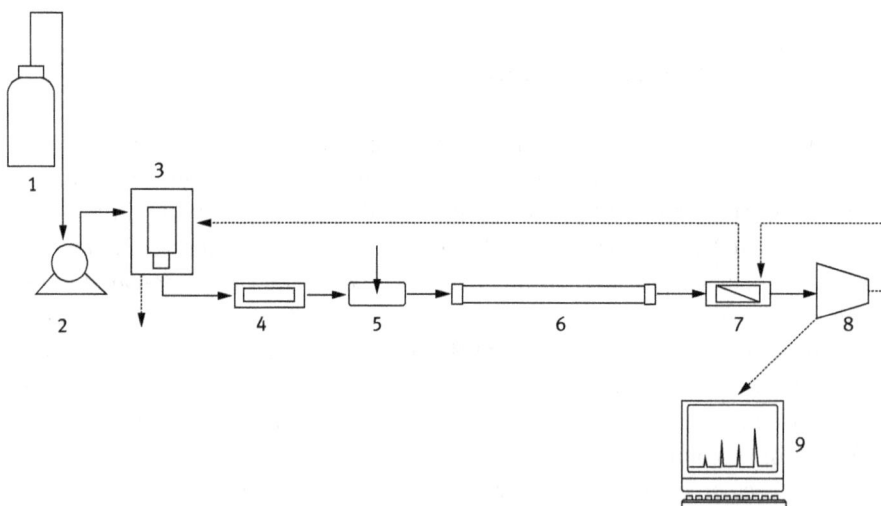

Figure 56: Schematic diagram of an RFIC system: 1, water reservoir; 2, pumping system; 3, eluent
generator; 4, continuously regenerated in-line trap column; 5, injector; 6, chromatographic
column; 7, suppressor; 8, detector; 9, computer.

In RFIC systems, the EG is placed between the high-pressure pump and the injector
(Figure 56); the pump delivers deionized water to an EG, where the eluent is electrolytically generated in-line. At the injector valve, the autosampler introduces the
sample through a sample loop to the chromatographic column, where the analytes
are separated and eluted into the suppressor (where the background conductivity is
reduced, in the meantime removing cations) then the analytes are detected.

Modern RFIC systems combine three technologies (Liu et al., 2007):

– electrolytic EG – which produces acid, base or salt eluents as required for IC
separations;

–   self-regenerating electrolytic suppression – produces the regenerant ions necessary for eluent suppression and allows continuous operation with less maintenance, and
–   continuously regenerated in-line trap column (CR-TC) – removes trace-level ionic impurities from the eluent.

Electrolytic suppressors improve the sensitivity and stability by reducing the conductivity of the eluent, while enhancing the conductivity of the analytes. The self-regenerating suppressor module is based on the return of the detector effluent and electrolysis to regenerate the suppressor, by recycling the mobile phase for use in the electrolytic generation of the regenerant acid (or base), eliminating the need for suppressor regeneration solutions.

CR-TC are a must in this context since contaminants will affect the purity and the concentration of the electrolytically generated eluents; these contaminants may cause additional peaks, increased background conductivity and even baseline shifts, compromising the separations, affecting reproducibility and detection limits.

Despite its relative novelty, there are numerous applications of RFIC, especially in the case of RFIC-MS systems, for the determination of anions, disinfection by products, haloacetic acids, bromate, pesticides, perchlorate, ethanolamines, detergents, bactericides, etc. (De Borba and Rohrer, 2004; Kim et al., 2008; Rohrer and DeBorba, 2007; Sheldon and Hoefler, 2008; Simsek et al., 2020; Thermo Fisher Scientific – Dionex Application Note 169; Yousef et al., 2012).

The major advantages of RFIC, outlined in most of applications, are:
–   It eliminates the need of manual eluent preparation, hence there are no chemicals to handle or reagents to prepare, while associated problems such as contamination and operator errors can be avoided (e.g., because the eluent is produced in-line, there is no possibility of contamination from carbonate).
–   It produces eluents with accurate and reproducible concentrations, leading to an enhanced retention time reproducibility.
–   Gradient separations are easy to achieve and highly reproducible, allowing the separation of mono- to polyvalent species in one run.
–   It provides a stable background and highly reproducible results, with little variability in peak retention times or areas.
–   Ease of use – classic eluent preparation is replaced by adding water; while the routine IC operation is simplified, the operator is free to focus on other analytical tasks rather than eluent preparation.
–   It reduces the overall operating costs.
–   It increases the sensitivity and the reliability of IC determinations.
–   It is compatible with the detectors used in IC.

Commercially available RFIC systems range from integrated ones (ICS-2000 being the world's first integrated RFIC system) to more complex modular ones, such as Dionex ICS-5000 + System Reagent-Free system.

4.2 Capillary ion chromatography

Capillary columns (columns having the internal diameter ranging from 0.1 to 0.5 mm) opened a new direction for IC – that of capillary IC (CIC). CIC can be described as an ion-exchange separation technique accomplished using flow rates in the range of several microliters per minute with capillary columns; since high backpressures are generated by using such columns, high-pressure systems is a must, hence the technique is also referred as high-pressure capillary ion chromatography (HPCIC or HPIC). Besides, using smaller particle column packings results in increase of efficiency, with narrower peaks, with a great impact on separation speed and resolution (the acronym HPIC also means high performance ion chromatography!); peak integration will also be improved, leading to more accurate and reliable results. Such systems provide improved resolution, speed, sensitivity and reproducibility; because of the very low flow rates, the eluent consumption is very small (up to 20 mL/day, 5–6 L/year), and such systems can be "always on" to analyze samples in any moment (Avdalovic and Liu, 2021; Iammarino et al., 2021; Kuban and Dasgupta, 2004; Liu et al., 2012; Sjögren et al, 1997).

Compared with the common IC systems, in HPIC the column size, the injection volumes and flow rates are scaled down by a factor ranging from 10 to 100 and the advantages they have are numerous, especially when these systems use automatic EG:
- provide a stable background with more consistent response with highly reproducible results, with little variability in peak retention times or areas;
- eliminates start-up and equilibration time;
- is remarkable fast;
- reduces the need for recalibration;
- reduces preventive maintenance and down time as well as the operating costs;
- when used in conjunction with an automatic eluent generator, they supplementary saves time by eliminating eluent preparation, start-up time and system equilibration;
- operating separations at low flow rates make them compatible with mass spectrometry detection;
- since the injection volumes are reduced to less than 0.5 µL, CIC is well suited for applications which are sample limited;
- overall, it can be considered as a "green" analytical alternative for the determination of most analytes, as it uses non-hazardous chemicals, generating a very small amount of non-toxic wastes.

The most appreciate issues are however that HPIC systems offer high separation efficiency while they have an increased productivity since they can operate continuously for an extended period of time, ensuring that they are always equilibrated and ready for sample analysis. Of course such advantages come with a price and HPIC systems are quite expensive because of their design adapted at very low flow rates, with micropumps, microscale suppressor, capillary trap columns, miniaturized EG and detectors (Yang et al., 2012; Thermo Fisher Scientific – capillary ion chromatography).

Compared with classical IC separations in which usually less than 10 peaks are obtained, in an HPIC separation, one can obtain ~30 peaks during 30–40 min, the peaks' profile being narrow and the chromatograms being similar with those obtained in GC. Using shorter column will improve resolution for less complex samples, while providing a faster analysis; hence, when samples have a low number of analytes, it can be more convenient to use shorter columns (150 mm instead of 250 mm); in such cases, by increasing the flow rate the duration of a separation can be decreased up to less than five minutes, hence increasing the productivity (Figure 57). The longer columns will help maximize resolution for more complex samples, but the maximum usable flow rate and therefore analysis speed is limited by the high back pressure (Kuban and Dasgupta, 2004; Yang et al., 2012).

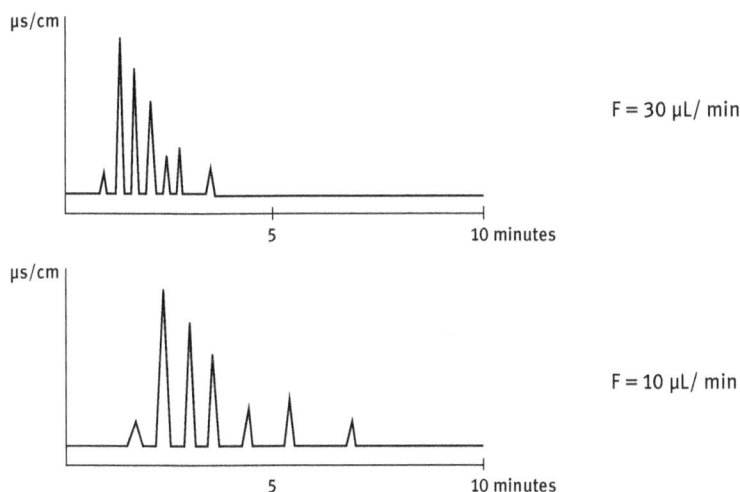

Figure 57: Reducing the separation time of anions from a standard mixture by increasing flow rates using a Dionex IonPac AS18-4 µm column. Peak ID: 1, fluoride; 2, chloride; 3, nitrite; 4, sulfate; 5, bromide; 6, nitrate; 7, phosphate (Dionex IonPac AS18 Fast 4 mm IC column).

As detection options, since all ions are electrically conductive, suppressed conductivity detection is the most common for the majority of CIC applications. Electrochemical detection is more selective and sensitive, being used in cases of compounds that contain an oxidizable or reducible moiety within their structure. Charge detection is a novel direction in CIC, being used in combination with conductivity detection to detect ions in proportion to their charge and concentration; ions with the same charge and concentration yield similar response (Dasgupta et al., 2013).

CIC is applicable in all the major fields of food analysis, extending the applicability of common IC in all cases (Table 15). Of particular interest is the analysis of beverages: lactate and acetate are important in evaluating the freshness, while HPIC fingerprints can help in differentiating various types of juices, being of real importance in establishing the products' authenticity (if certain adulteration practices are suspected). HPIC fingerprinting is an important tool not only in the authentication of juices but also of coffee, wine or honey (Choi et al., 2012; Dzolin et al., 2019; Fermo et al, 2013; Kelly et al., 2005; Yashin et al., 2017).

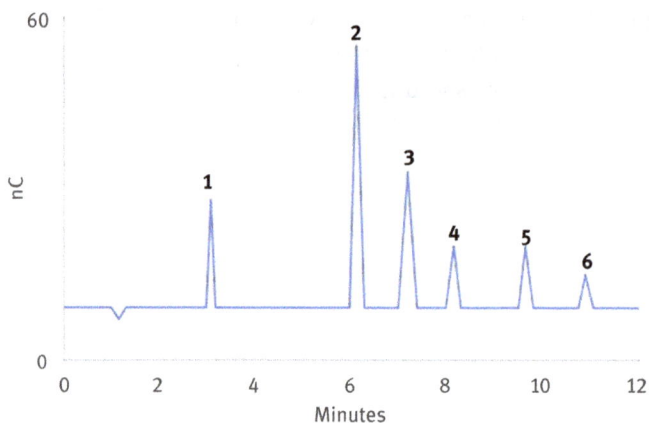

Figure 58: Separation of carbohydrates from beverages with a CarboPac PA20 column using 10 mM KOH (EG). Peak ID: 1, fucose; 2, galactosamine; 3, glucosamine; 4, galactose; 5, glucose; 6, mannose (Christison et al., 2016c).

If the separations reported for preservatives, nitrites, nitrates, major anions and major cations from different food products are not so spectacular, being remarkable eventually by a very short duration (Figures 58, 59 and 60), certain separations of organic acids by CIC are outstanding both by the high number of analytes and short duration (e.g., Doyle, 2016; Yang et al., 2016).

Table 15: Summary of CIC applications in food analysis.

Analytes/matrix	Column	Mobile phase	Detection	Separation ltime (min)	Reference
Preservatives (benzoic acid, sorbic acid)/food products	IonPac® AS11-HC (250 × 0.4 mm), 9 μm	EGC KOH – gradient 0.025 mL/min	Suppressed conductivity	35	D'Ámore et al. (2021)
Nitrite and nitrate/meat products	IonPac® AS11-HC (250 × 0.4 mm), 9 μm	EGC KOH – gradient 0.015 mL/min	Suppressed conductivity	20	D'Ámore et al. (2019)
Organic acids and inorganic anions (29 analytes)/orange juice	IonPac® AS11-HC (250 × 0.4 mm), 9 μm	EGC KOH – gradient 0.015 mL/min	Suppressed conductivity	36	Doyle M. (2016)
Organic acids and inorganic anions (15 analytes)/orange juice – spoilage identification	IonPac® AS11-HC (250 × 0.4 mm), 9 μm	EGC KOH – gradient 0.015 mL/min	Suppressed conductivity	36	Doyle M. (2016)
)Organic acids and inorganic anions (25 analytes)/apple and cranberry juices	IonPac® AG11-HC/AS11-HC (250 × 0.4 mm), 9 μm	EGC KOH – gradient 0.015 mL/min	Suppressed conductivity	45	Doyle M. (2016)
Organic acids and inorganic anions (14 analytes)/beer	IonPac® AG11-HC/AS11-HC (250 × 0.4 mm), 9 μm	EGC KOH – gradient 0.015 mL/min	Suppressed conductivity	40	Doyle M. (2016)
Organic acids and inorganic anions (13 analytes)/apple and orange juices	IonSwift MAX 100 (250 × 0.25 mm)	EGC KOH – gradient 0.015 mL/min	Suppressed conductivity	20	Thermo Fisher Scientific Application Brief 137
Quinate, glycolate, chloride, malate, sulfate, maleate, phosphate/young coconut water	IonSwift MAX 100 (250 × 0.25 mm)	EGC KOH – gradient 0.012 mL/min	Suppressed conductivity	22	Thermo Fisher Scientific, 2011a

(continued)

Table 15 (continued)

Analytes/matrix	Column	Mobile phase	Detection	Separation (time (min)	Reference
Organic acids, inorganic anions (incl. arsenate, 27 analytes)/fruit juices	IonPac AS11-HC-4 µm (250 × 0.4 mm)	EGC KOH – gradient 0.015 mL/min	Suppressed conductivity	40	Yang et al. (2016)
Fucose, sucrose, arabinose, galactose, glucose, xylose, mannose, fructose/corn stover	CarboPac SA10 (250 × 2 mm), 4 µm	EGC KOH – gradient 0.038 mL/ min	Pulsed amperometric detection	6	Doyle M. (2016)
Fucose, galactosamine, glucosamine, galactose, glucose, manose/beverages	CarboPac PA20 (150 × 0.4 mm)	EGC KOH – 10 mM KOH 0.008 mL/min	Pulsed amperometric detection	12	Christison et al. (2016c)
Mannitol, galactosamine, arabinose, galactose, glucose, sucrose, xylose, mannose, fructose, ribose/coffee	CarboPac PA20, (150 × 0.4 mm)	EGC KOH – gradient 0.09 mL/min	Pulsed amperometric detection	45	Liu et al. (2012)
Chloride, nitrate, sulfate, phosphate, citrate/beverages	IonSwift MAX 200 (250 × 0.25 mm)	EGC KOH – gradient 0.025 mL/min	Suppressed conductivity	8	Christison et al. (2016a)
Chloride, nitrate, sulfate, phosphate, citrate/beverages	IonSwift MAX 100 (250 × 0.25 mm)	EGC KOH – gradient 0.024 mL/min	Suppressed conductivity	12	Christison et al. (2016b)
Sodium, potassium, magnesium and calcium/young coconut water	IonPac CG12A, CS12 (250 × 0.25 mm)	35 mM methanesulfonic acid 0.012 mL/ min	Suppressed conductivity	20	Thermo Fisher Scientific, 2011b

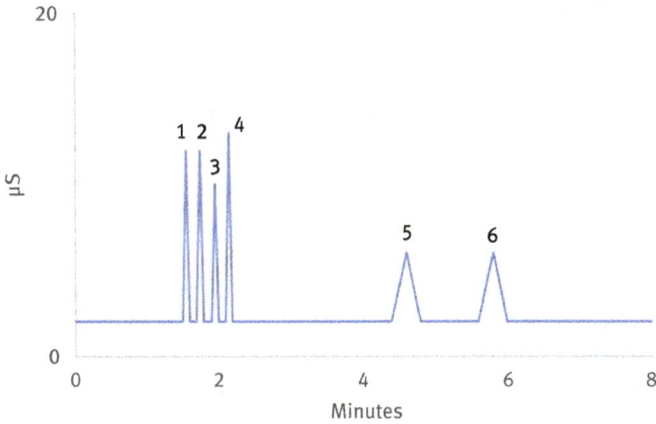

Figure 59: Fast separation of cations from water on a capillary Ion Pac CS12A-5 µm column using 20 mM methane sulfonic acid (EG) and 18 µL/min flow rate. Peak ID: 1, lithium; 2, sodium; 3, potassium; 4, ammonium; 5, magnesium; 6, calcium (Ion Pac CS12A Cation – Exchange Column).

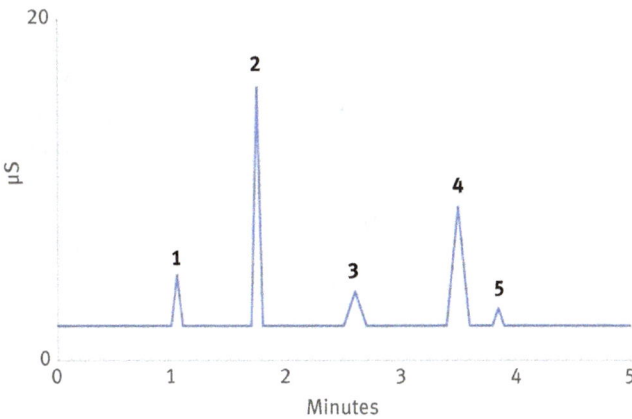

Figure 60: Rapid separation of major anions from drinking water using Dionex IonPac AG18-Fast column/AS 18 Fast column, 23 mM KOH (EGC), 0.55 mL/min/suppressed conductivity. Peak ID: 1, fluoride; 2, chloride; 3, carbonate; 4, sulfate; 5, nitrate (Dionex IonPac AS18-Fast-4 µm IC columns).

4.3 Combustion ion chromatography

Combustion ion chromatography (CoIC) is a version of IC which integrates inline combustion sample preparation, enabling the simultaneous determination of halogens (as fluoride, bromide, chloride and iodide) and sulphur in an automated process; the products of combustion are adsorbed in hydrogen peroxide or in ultrapure water and then analyzed through IC (Emmegger et al., 2010; Lauebli et al., 2022).

CoIC allows hence a fast halogen and sulphur analysis in a single run, by combining sample preparation and analysis in one step; the sample (solid or liquid) is placed in a sample boat located in an autosampler, from which it is loaded into a combustion oven where pyrolysis occurs, followed by oxidation in an O_2 and water vapor stream, at temperatures above 900°C. Combustion breaks the bounds from organic compounds releasing SOx and hydrogen halides or gaseous halogens, then the resulted gases are transported into an absorption unit, where they react further with H_2O_2, sulfur oxides being oxidized to produce finally sulfate and halogens form halide ions (Figure 61). An aliquot from the resulting solution containing the anions of interest (halides and sulfate) is finally subjected to IC analysis (Emmenegger et al., 2010).

The IC separation is quite a simple one, being accomplished with an anion-exchange column, followed by a suppressor and a conductivity detector, the elution order for the analytes being: $F > Cl > Br > SO_4^{2-}$. Since the concentration of the analytes is very low, suppression is necessary in this context to reduce the background conductivity and to enhance the analytes' signals.

The overall duration of a combustion IC analysis is around 15 min, being determined by the IC separation time; in automatic systems, in the meantime with IC separation, pyrolysis of the next sample can be accomplished, hence this analytical approach is time-efficient, providing the desired results in a short time.

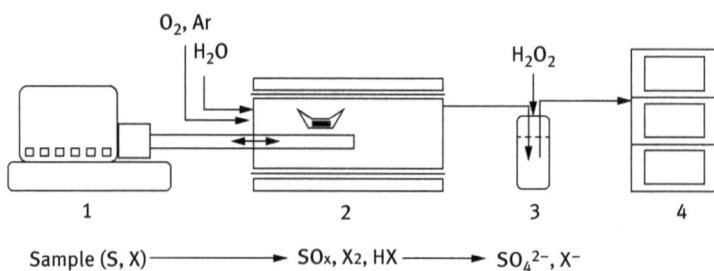

Figure 61: Schematic flowchart diagram for the combustion IC process: 1, sample charger; 2, furnace; 3, absorption system; 4, ion chromatograph (Yang et al., 2017).

Being so convenient, CoIC can be used for a broad range of applications in petroleum, coal, plastic, semiconductor, pharmaceutical, environmental, extractive and power generation industries (von Abercron et al., 2019; ASTM D7359-18; Hemmler et al., 2014; Li et al., 2021; Michalski, 2014; Miyake et al., 2007). The success of this method is because the determination of the halogen and sulfur content in complex matrices is a challenge, mainly due to the fact that the sample preparation is demanding, highly time-consuming and error-prone; such drawbacks can be overcome by CoIC, which eliminates complex sample preparation using an easy-to-use, highly sensitive automated method, saves time and produces fewer environmental contaminants than other sample preparation procedures.

Because CoIC is a new entry on the market of analytical techniques, up to now there is no accredited method involving it in food analysis. For the same reason, there are not many references for implementation of CoIC: determination of brominated vegetable oils in soft drinks was reported (Christison et al., 2019; Thermo Fischer Scientific Application Note 72,917) as well as the determination of fluoride in tea (Yang et al., 2017), determination of fluorine in Krill oil (Jung et al., 2020), determination of fluorine in food packages (Schultes et al., 2019) or the determination of halogens and sulfur in honey (Mesko et al., 2020).

An important disadvantage of CoIC is related to the determination of inorganic-bonded halides and sulfur, these yielding poor recoveries, since the temperature-limit of pyrolysis is 1100 °C, while inorganic halides have higher boiling points; this method is more appropriate for the determination of organic-bounded sulfur and halogens.

CoIC systems can be acquired from Thermo Fisher Scientific, Nittoseiko Analytech (Mitsubishi AQF-2100H automatic quick furnace for CoIC), Metrohm or AE-MIC Trading (AE-MIC Trading, Nittoseiko Analytech, Metrohm – CoIC; Thermo Fisher Scientific – combustion ion chromatography system).

The systems are modular, consisting in separate units for each operation; for example, Metrohm offers the following configuration:
- an autosampler module, which transfers the solid or liquid samples into the combustion module, where pyrolysis takes place;
- an absorber module, which traps the gases resulted from pyrolysis;
- a 930 IC compact (for IC analysis).

References

AE-MIC Trading. Automatic combustion ion chromatography analysis system for halogens and sulfur. (Accessed April 19, 2022, at http://ae-mic-trading.com/cic).
ASTM D7359-18 Standard test method for total fluorine, chlorine and sulfur in aromatic hydrocarbons and their mixtures by oxidative pyrohydrolytic combustion followed by ion chromatography detection (Combustion Ion Chromatography-CIC) (Accessed April 19, 2022, at https://www.astm.org/d7359-18.html)

Avdalovic, N., & Liu, Y. (2021). Capillary ion chromatography. In Separation Science and Technology. Academic Press, Vol. 13, 303–322.

Choi, J. Y., Bang, K. H., Han, K. Y., & Noh, B. S. (2012). Discrimination analysis of the geographical origin of foods. Korean Journal of Food Science and Technology, 44(5), 503–525.

Christison, T., Ellison, G., & Rohrer, J. (2019) Determination of brominated vegetable oils in soft drinks using combustion ion chromatography. Thermo Fisher Scientific, Sunnyvale, CA. (Accessed April 19, 2022, at https://assets.thermofisher.com/TFS-Assets/CMD/posters/po-90510-ic-brominated-vegetable-oils-soft-drinks-rafa2019-po90510-en.pdf)

Christison, T., Pang, F., & Lopez, L. (2016a). Fast separations of anions and organic acids in a carbonated beverage using high-pressure capillary IC – Technical Note 118, Thermo Fisher Scientific, Sunnyvale, CA, USA. (Accessed April 11, 2022, at https://www.thermoscientific.com/content/dam/tfs/ATG/CMD/CMD%20Documents/TN-118-Fast-Separations-Anions-Organic-Acids-Carbonated-Beverage-TN70167-E.pdf)

Christison, T., Pang, F., & Lopez, L. (2016b). Fast separations of organic acids in an orange juice sample using high-pressure capillary IC – Technical Note 119, Thermo Fisher Scientific, Sunnyvale, CA, USA. (Accessed April 11, 2022, at https://assets.thermofisher.com/TFS-Assets/CMD/Technical-Notes/tn-119-hpic-organic-acids-orange-juice-tn70168-en.pdf).

Christison, T., Zhang, A., & Lopez, L. (2016c). Determinations of monosaccharides and disaccharides in beverages by capillary HPAE-PAD- Technical Note 135, Thermo Fisher Scientific, Sunnyvale, CA, USA. (Accessed April 11, 2022, at https://www.thermoscientific.com/content/dam/tfs/ATG/CMD/CMD%20Documents/Application%20&%20Technical%20Notes/Chromatography/Ion%20Chromatography/IC%20and%20RFIC%20Systems/TN-135-Determination-Monosaccharides-Disaccharides-Beverages-TN70646-E.pdf)

D'Amore, T., Di Taranto, A., Vita, V., Berardi, G., & Iammarino, M. (2019). Development and validation of an analytical method for nitrite and nitrate determination in meat products by capillary ion chromatography. Food Analytical Methods, 12(8), 1813–1822.

D'Amore, T., Di Taranto, A., Berardi, G., Vita, V., & Iammarino, M. (2021). Going green in food analysis: A rapid and accurate method for the determination of sorbic acid and benzoic acid in foods by capillary ion chromatography with conductivity detection. Lwt, 141, 110841,

Dasgupta, P. K., Liao, H., & Shelor, C. P. (2013). Ion chromatography yesterday and today: Detection. LC-GC North America, 31(4B), 23–26.

De Borba, B., & Rohrer, J. (2004). Determination of benzoate in liquid food products by reagent-free ion chromatography. LC-GC North America, 22, 39–39,

Dionex IonPac AS18-Fast-4μm IC Columns, (Accessed April 11, 2022, at https://www.thermofisher.com/order/catalog/product/076033)

Doyle, M. (2016). High-pressure ion chromatography, applications in food and beverage analysis. In "Advancing ion chromatography with high pressure. (Accessed April 11, 2022, at https://assets.thermofisher.com/TFS-Assets/CMD/brochures/EB-IC-Ion-Chromatography-High-Pressure-EN.pdf).

Dzolin, S., Ibrahim, W. A. W., Mahat, N. A., Keyon, A. S. A., & Ismail, Z. (2019). Unique signatures of honeys as a means to establish provenance. Malaysian Journal of Analytical Sciences, 23(1), 1–13.

Emmenegger, C., Wille, A., & Steinbach, A. (2010). Sulphur and halide determination by combustion ion chromatography. LC-GC. (Accessed April 19, 2022, at https://www.chromatographyonline.com/view/sulphur-and-halide-determination-combustion-ion-chromatography).

Fermo, P., Beretta, G., Facino, R. M., Gelmini, F., & Piazzalunga, A. (2013). Ionic profile of honey as a potential indicator of botanical origin and global environmental pollution. Environmental Pollution, 178, 173–181,

Haddad, P. R., Nesterenko, P. N., & Buchberger, W. (2008). Recent developments and emerging directions in ion chromatography. Journal of Chromatography. A, 1184(1–2), 456–473.

Hemmler, D., Emmenegger, C., Schmitz, D., Kaufmann, S., & Steinbach, A. (2014). Determination of halogens and sulphur in complex matrices. LC-GC, (Accessed April 19, 2022, at https://www.chromatographyonline.com/view/determination-halogens-and-sulphur-complex-matrices).

Iammarino, M., Berardi, G., D'Amore, T., Vita, V., & Di Tarant, A. (2021) Exploring the potentiality of capillary ion chromatography (CIC) as analytical technique for the determination of food additives. Book of abstracts – XXVII Congresso Nazionale della Società Chimica Italiana "La chimica guida lo sviluppo sostenibile" – 14-23.09. 2021, ANA ORO31.

IonPac CS12A Cation – Exchange Column, Thermo Scientific/ Dionex (Accessed April 11, 2022, at https://tools.thermofisher.com/content/sfs/brochures/4230-DS-IonPac-CS12A-20Apr2010-0643-12-R2.pdf)

Jung, J., Kim, S., Chu, E., & Joung, J. (2020). Determination of fluorine in Krill oils by combustion-ion chromatography. Analytical Science and Technology, 33(6), 262–273.

Kelly, S., Heaton, K., & Hoogewerff, J. (2005). Tracing the geographical origin of food: The application of multi-element and multi-isotope analysis. Trends in Food Science & Technology, 16(12), 555–567.

Kim, H. B., Sim, W. J., Kim, M. Y., & Oh, J. E. (2008). Monitoring and evaluation of analytical methods of perchlorate with IC and LC/MS. Journal of Korean Society of Environmental Engineers, 30(1), 37–44.

Kuban, P., & Dasgupta, P. K. (2004). Capillary ion chromatography. Journal of Separation Science, 27(17/18), 1441–1457.

Laeubli, M., Emmenegger, C., & Bogenschütz, G. Ion chromatographic determination of halogens and sulfur in solids using combustion as inline sample preparation. Metrohm. (Accessed April 19, 2022, at https://partners.metrohm.com/GetDocumentPublic?action=get_dms_document&docid=1552200)

Li, T., Min, H., Li, C., Yan, C., Zhang, L., & Liu, S. (2021). Simultaneous determination of trace fluorine and chlorine in iron ore by combustion-ion chromatography. Analytical Letters, 54(15), 2498–2508.

Liu, Y., Barreto, V., Cheng, J., Jandik, P., & Pohl, C. (2012). New developments in capillary ion chromatography systems using on-line electrolytic eluent generation – white paper 70461 Thermo Fisher Scientific, Sunnyvale, CA, USA. (Accessed April 11, 2022, at https://tools.thermofisher.com/content/sfs/brochures/WP-70461-New-Developments-Capillary-IC-Systems.pdf)

Liu, Y., Lu, Z., Pohl, C., Madden, J., & Shirakawa, N. (2007). Reagent-free ion chromatography systems with eluent regeneration: RFIC-ER systems. American Laboratory, 39(3), 17–19.

Mesko, M. F., Balbinot, F. P., Scaglioni, P. T., Nascimento, M. S., Picoloto, R. S., & da Costa, V. C. (2020). Determination of halogens and sulfur in honey: A green analytical method using a single analysis. Analytical and Bioanalytical Chemistry, 412(24), 6475–6484.

Metrohm – combustion ion chromatography. (Accessed April 19, 2022, at https://partners.metrohm.com/GetDocumentPublic?action=get_dms_document&docid=2991258).

Michalski, R. (2014). Recent development and applications of ion chromatography. Current Chromatography, 1(2), 90–99.

Miyake, Y., Kato, M., & Urano, K. (2007). A method for measuring semi-and non-volatile organic halogens by combustion ion chromatography. Journal of Chromatography. A, 1139(1), 63–69.

Nittoseiko Analytech – AQF-2100H Automatic quick furnace for combustion ion chromatography. (Accessed April 19, 2022, at https://www.n-analytech.co.jp/global/instrument/c-ic/entry-173.html)

Rohrer, J., & De Borba, B. (2007). Determination of trace concentrations of bromate and bromide in natural mineral waters by reagent-free ion chromatography. LCGC Asia Pacific, 10(2), 40–40.

Schultes, L., Peaslee, G. F., Brockman, J. D., Majumdar, A., McGuinness, S. R., Wilkinson, J. T., & Benskin, J. P. (2019). Total fluorine measurements in food packaging: How do current methods perform?. Environmental Science & Technology Letters, 6(2), 73–78.

Sheldon, B., & Hoefler, F. (2008). New ion chromatography solutions for routine water analysis. American Laboratory, 40(19), 12–14.

Şimşek, M. G., Destanoğlu, O., & Yilmaz, G. G. (2020). Simultaneous determination of fluoride, acetate, formate, chloride, nitrate and sulfate in distilled alcoholic beverages with ion chromatography/conductivity detector. Journal of the Turkish Chemical Society Section A: Chemistry, 7(3), 661–674.

Sjögren, A., Boring, C. B., Dasgupta, P. K., & Alexander IV, J. N. (1997). Capillary ion chromatography with on-line high-pressure electrodialytic NaOH eluent production and gradient generation. Analytical Chemistry, 69(7), 1385–1391.

Srinivasan, K. (2021). Electrolysis-based accessories in ion chromatography. In Separation Science and Technology. Academic Press, Vol. 13, 203–218.

Strong, D. L., Dasgupta, P. K., Friedman, K., & Stillian, J. R. (1991). Electrodialytic eluent production and gradient generation in ion chromatography. Analytical Chemistry, 63(5), 480–486.

Thermo Fisher Scientific (2011a). Anion determinations in a young coconut water sample on a Dionex IonSwift MAX-100 capillary column. (Accessed April 11, 2022, at https://tools.thermo fisher.com/content/sfs/brochures/110969-28876-anion-organic-acid-coconut-water-IonSwift-MAX-100.pdf)

Thermo Fisher Scientific (2011b). Cation determinations of a young coconut water sample on a Dionex IonPac CS12A capillary column. (Accessed April 11, 2022, at https://tools.thermofisher.com/con tent/sfs/brochures/110968-28875-inorganic-cation-coconut-water-IonPac-CS12A.pdf).

Thermo Fisher Scientific Application Brief 137 (2015). Determination of inorganic anions and organic acids in apple and orange juice samples using capillary IC, Dionex LPN 2970, Sunnyvale, CA. (Accessed April 11, 2022, at https://appslab.thermofisher.com/App/2387/or ganic-acids-apple-orange-juice)

Thermo Fisher Scientific. Dionex Application Note 169, Rapid determination of phosphate and citrate in carbonated soft drinks using a reagent-free ion chromatography, Dionex LPN 1774. Sunnyvale, CA, 2005. (Accessed July 17, 2021, at https://assets.thermofisher.com/TFS-Assets /CMD/Application-Notes/AN-169-IC-Phosphate-Citrate-Soft-Drinks-AN71409-EN.pdf)

Thermo Fischer Scientific Application Note 72917 Fast determinations of brominated compounds in carbonated beverages using oxidative pyrolytic combustion and ion chromatography. (Accessed April 19, 2022, at https://assets.thermofisher.com/TFS-Assets/CMD/Application-Notes/an-72917-ic-brominated-compounds-beverages-an72917.pdf)

Thermo Fisher Scientific – capillary ion chromatography. (Accessed April 11, 2022, at https://www.thermofisher.com/ro/en/home/industrial/chromatography/chromatography-learning-center /ion-chromatography-information/ion-chromatography-innovations/capillary-ion-chromatography.html)

Thermo Fisher Scientific – combustion ion chromatography system. (Accessed April 19, 2022, at https://www.thermofisher.com/order/catalog/product/IQLAAAGADHFAMJMBIQ).

von Abercron, E., Falk, S., Stahl, T., Georgii, S., Hamscher, G., Brunn, H., & Schmitz, F. (2019). Determination of adsorbable organically bound fluorine and adsorbable organically bound halogens as sum parameters in aqueous environmental samples using combustion ion chromatography. Science of the Total Environment, 673, 384–391,

Yang, B., Zhang, F., & Liang, X. (2012). Recent development in capillary ion chromatography technology. Open Chemistry, 10(3), 472–479.

Yang, H., Hu, J., & Rohrer, J. (2017) Determination of fluoride in tea using a combustion ion chromatography system. Thermo Scientific Application Note AN72268, Thermo Fisher Scientific,

Sunnyvale, CA. (Accessed April 19, 2022, at https://tools.thermofisher.com/content/sfs/bro chures/AN-72268-IC-Fluoride-Tea-AN72268-EN.pdf and at https://assets.thermofisher.com/TFS-Assets/CMD/posters/PO-72498-IC-CIC-Fluoride-Tea-AOAC2017-PO72498-EN.pdf)

Yang, H., Christison, T., & Lopez, L. (2016), Determination of total inorganic arsenic in fruit juice using high-pressure capillary ion chromatography – Technical Note 145, Thermo Fisher Scientific, Sunnyvale, CA, USA. (Accessed April 11, 2022, at https://assets.thermofisher.com/TFS-Assets /CMD/Technical-Notes/tn-145-hpic-total-inorganic-arsenic-fruit-juice-tn70881-en.pdf).

Yashin, A., Yashin, Y., Xia, X., & Nemzer, B. (2017). Chromatographic methods for coffee analysis: A review. Journal of Food Research, 6(4), 60–82.

Yousef, A. A., Abbas, A. B., Badawi, B. S., Al-Jowhar, W. Y., Zain, E. A., & El-Mufti, S. A. (2012). Rapid quantitative method for total brominated vegetable oil in soft drinks using ion chromatography. Food Additives & Contaminants: Part A, 29(8), 1239–1243.

5 Sample preparation for ion chromatography

To be analyzed by IC, an ideal sample should be a clear aqueous solution, containing no interfering compounds, no particulate matter and no bacteria, while having an appropriate concentration and a pH close to 7. Real samples are more or less far from this ideal: solids, gels, emulsions, oils or particle-containing systems are common ones. Certain samples contain the analytes in very high concentrations, while others contain trace levels of substances of interest; some samples are very acidic, others are highly basic (Haddad et al., 1999). There are numerous cases in which food samples contain important amounts of interfering substances that can make the analysis very difficult, disturbing the chromatographic process (e.g., by causing changes in retention times, baseline issues, decreasing resolution and efficiency) and leading to poor reproducibility and accuracy. In such cases, a preliminary sample preparation stage is a must before the IC analysis for certain categories of samples (e.g., milk, cream, emulsions, chocolate and wastewater).

In such a context, the major objectives of sample preparation are (Lundanes et al., 2013; Moldoveanu and David, 2021; Pomeranz, 2013; Slingsby and Kiser, 2001):
- to adapt the analyte's concentration at the linearity range of the method (by dilution or concentration);
- to clean up the sample matrix;
- to achieve analytical compatibility with the IC system.

Numerous sample preparation methods are nowadays available, and their choice depends on the aggregation state and the composition of the sample:
- for solid samples, the simplest is dissolving, but when the analytes are in a complex matrix, solid-liquid extraction is necessary, being usually accomplished providing proper homogenization, in some cases assisted by sonication and heat/microwaves; the most difficult solids are those requiring disruption, either using acids or bases, or enzymes;
- for liquid samples, direct analysis is the simplest approach but even in such a happy situation, filtration is necessary to remove unwanted particles; in numerous cases, a dilution or concentration stage is necessary as well as a pH adjustment; for more complex samples, one can apply common methods such as solid-phase extraction (SPE), dialysis or ultrafiltration.

Typically sample preparation for IC includes dilution and/or filtering through 0.45 μm membrane filters; under certain circumstances, samples may require removal of undesirable species through SPE techniques. Manual sample preparation is common, as this is simpler and cheaper; unfortunately, such an approach is time-consuming and can give rise to poor reproducibility issues and inaccurate results, but a higher budget allocated to the analytical laboratory allows the acquisition of automatic sample

https://doi.org/10.1515/9783110644401-005

processors or of an in-line sample preparation technique. There are a large array of commercially available in-line sample preparation solutions, such as ultrafiltration, dialysis, preconcentration, cation removal, dilution, extraction, matrix elimination and neutralization (MISP). Using in-line sample preparation is preferable, given the numerous advantages of automated sample preparation: improved precision and accuracy of the analytical results, reduced processing time, reduced manual work, traceability for each sample preparation step.

5.1 Dilution

Dilution is the process of decreasing the concentration of an analyte from a sample, usually by adding more water to the sample. Usually, food samples contain several analytes at really high concentrations, hence they have to be diluted in order to be analyzed. Knowing that IC is appropriate for analysis in the range of around 1 µg/ L to 100 mg/L, the first step in the analysis is to ensure that the sample has the appropriate concentration to be analyzed. In fact, the lower the concentrations are, the better the accuracy of the results. A proper dilution can also help in avoiding overloading problems as well as of possible interferences caused by certain analytes in very high concentrations. Working at lower concentrations of the analytes is also advantageous since it can increase significantly the lifetime of IC columns.

Sample dilution is usually accomplished by pipetting a certain volume of sample solution into a volumetric flask and then filling the flask to the mark with distilled or ultrapure water; another choice is with the mobile phase used in the IC analysis.

Because concentration is defined as quantity of compound/volume of sample solution, expanding the volume of the sample solution will decrease the concentration of all compounds in the original sample. To obtain the concentration of a certain analyte after a dilution was accomplished, one has to multiply the corresponding reported result with the dilution factor. It is possible to use the formula:

$$C_{in} \times V_{in} = C_{dil} \times V_{dil}$$

from which

$$C_{in} = C_{dil} \times \frac{V_{dil}}{V_{in}} = C_{dil} \times DF$$

where:
C_{in} – the initial concentration of the analyte in the sample (before dilution);
C_{dil} – the final concentration of the analyte in the sample (after dilution);
V_{in} – the initial volume of the sample (before dilution);
V_{dil} – the final volume of the sample (after dilution);
DF – the dilution factor.

In-line dilution is an useful approach which can save time when dealing with samples which differs much in concentrations; when a sample with the analyte(s) concentration(s) outside the calibration range is detected, the chromatographic data system can decide and accomplish the necessary dilution after which the sample is re-injected. In-line dilution can also help in the calibration procedure, in which a mixture of standards at a given concentration can be diluted automatically in order to cover the desired range of concentrations.

5.2 Filtration

Filtration removes particulate matter from the sample solution, protecting thus the injection valve components, the pumping system as well as the chromatographic column. Sample filtration prior to injection is a must because after injection the sample is carried to the chromatographic system and particulate matter from the sample can damage the injection valve, block the tubing and/or the chromatographic column, leading to irregularities in flow rate which can shift peaks' retention times, also increasing the backpressure. In severe cases, the columns are compromised and need to be replaced (Fan and Zhu, 2007; Frenzel and Markeviciute, 2017; Liu et al., 2015).

Sample filtration for IC is carried out using special single-use membrane filters (Figure 62), attached to plastic syringes; generally, 0.47-μm pore size membrane filters are appropriate, but small diameter capillary columns packed with particles under 3 μm will require 0.2 μm pore size filters. A wide range of commercial filters are available such as nylon, polytetrafluoroethylene, polyvinyldienedifluoride, or methylcellulose-based membrane filters; their choice depends on the characteristics of the samples and on the pore diameters of the membrane.

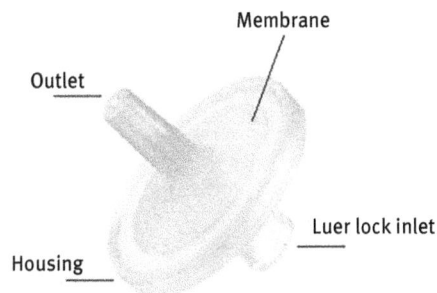

Figure 62: Membrane filter.

For filtration, a sample has to be aspired into a syringe, then the membrane filter is attached to the syringe and the syringe piston is gentle pushed; the filtrate is collected into a sample vial, which is finally capped and labeled. When performing membrane filtration with commercial disposable filters, one has to consider the risk of possible contamination and/or the loss of analyte by adsorption on the filter, hence it is recommended to reject the first milliliter of the filtrate to waste. A more convenient way to accomplish filtration is by using high-pressure in-line filters, mounted before the guard column (see Section 2.1.6).

Sample filtration will not only improve the quality of results but will also extend the columns' lifetime and reduce maintenance costs and time for the system components (Lundanes et al., 2013).

5.3 Ultrafiltration

Ultrafiltration is a membrane technique in which pressure or a concentration gradient leads to the separation of heterogeneous disperse systems through a semipermeable membrane (Frenzel and Markeviciute, 2017; Mohammad et al., 2012; Wiśniewski, 2019). The particles in suspension or compounds with high molecular weight are retained (in "retenate"), while water and low molecular weight solutes pass through the membrane in the filtrate ("permeate"). In-line ultrafiltration is especially useful, being a fast alternative to manual filtration, which does not require a separate sample preparation stage, taking place during IC analysis; after injection, particles from the samples remain on one side of the cellulose membrane, while analyte(s) pass through the membrane. Ultrafiltration cells have two compartments, separated by a filter membrane (Figure 63); on one side, the sample is carried at a high flow rate, while on the other side, some of the sample components are drawn off through the membrane and transported to the injection valve. Membrane clogging will not occur in such cells because of the continuous flow on the sample side, particulate matter being flushed away (MISP).

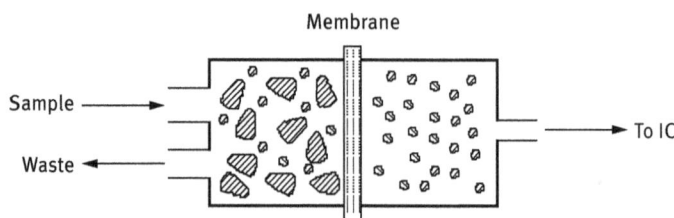

Figure 63: Schematic of an in-line ultrafiltration cell.

Ultrafiltration can be used in cases of samples containing moderate amounts of particles, various extracts, diluted fruit or vegetable juices, process water, wastewater, and so on.

As major improvements offered by in-line ultrafiltration, one can mention:
- there is no need to filter the samples;
- increase the lifetime of the column;
- one membrane is sufficient for multiple determination (e.g., Metrohm's systems can use a membrane for ~100 samples);
- time savings;
- lower operational costs.

5.4 Dialysis

Dialysis is a method for separation of low-molecular weight compounds from high-molecular weight compounds by means of a semipermeable membrane (Campbell, 2012; Thomas et al., 2021). In sample preparation for IC, dialysis is very useful in cases of samples with high organic load because certain organic molecules can contaminate the chromatographic column and disturb the separations (matrix effect). This technique is able to separate the target analytes from complex samples containing not only particles but also colloids, oil components and large molecules for which, in most cases, classic filtration is not applicable to get rid of these compounds.

Like ultrafiltration, dialysis can also be accomplished in-line; in-line dialysis ensures the separation of not only particulate matter but also of colloids and of large molecules (e.g., proteins), enabling a simpler analysis for a wide array of food samples, such as emulsions, dispersions, waste water, fermentation broths, milk, dairy products, concentrated fruit and vegetable juices, samples heavily contaminated with particles, algae or bacteria (Rick et al., 2009).

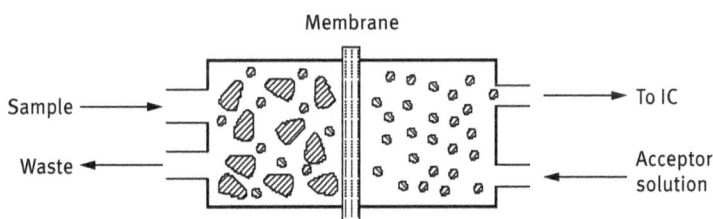

Figure 64: Schematic of an stop-flow in-line dialysis cell.

In-line dialysis stopped-flow cells have two compartments, separated by a cellulose acetate or nylon membrane (Figure 64). Samples are pumped on the sample side, and after a rinsing phase, the acceptor stream to the right of the membrane is halted. The ions pass through the membrane because of the concentration gradient, until an equilibrium is achieved (when the concentration in the acceptor solution matches the concentration of the original sample), then the acceptor solution is injected into the IC system; one membrane is sufficient for multiple determinations (MISP).

In-line dialysis is advantageous (Buldini et al., 2000; De Borba et al., 2001; Frenzel and Markeviciute, 2017; Gandhi, 2003; Nordmeyer and Hansen, 1982) since it enables the analysis of samples with high amounts of organic substances, increases the lifetime of the column and provides important time savings (e.g., samples containing proteins can be injected directly when using inline dialysis, avoiding time-consuming manual steps such as deproteinization with Carrez reagent). In food analysis, dialysis was reported in a wide area of applications, including the analysis of bromate in flour (Song et al. 2006), nitrate and nitrite in cooked meat (Yao et al., 2010), choline in milk powder (Li et al., 2008), carbohydrates and organic acids in beverages (Vérette et al., 1995).

5.5 Solid-phase extraction

SPE uses a small amount of solid packing material in a cartridge to retain selectively the analytes, enabling the concentration and purification of samples in a simple and convenient way (Simpson, 2000; Żwir-Ferenc and Biziuk, 2006). In principle, a sample solution is passed through a preconditioned SPE cartridge; the solid packing (strong ion exchanger for retaining ionic species, reversed-phase packing for retaining hydrophobic components) can either trap selectively the compounds of interest (which can be eluted later by using a small volume of an appropriate eluent) or retain the unwanted components, leaving ionic analytes elute. Because the gravitational fluid passage through the stationary phase is too slow, it is necessary to ensure either vacuum or a positive pressure, but usually SPE cartridges are used by fitting them to a vacuum manifold connected with a vacuum pump (Buszewski and, 2012; Poole, 2002).

When selecting a certain SPE cartridge, it is important to define the retention objective (Fritz, 1999):

- if the retention objective is to retain the contaminants, then the desired analyte (s) passes through the cartridge;
- if the retention objective is to retain the desired analyte(s), then the contaminants will be washed off and the target substances are eluted later for analysis.

If the contaminants are retained, the working procedure involves preliminary conditioning of the cartridge (with ultrapure water or low-strength buffer), then the sample is applied to the top of the cartridge using ~1 mL/min flow rate and the eluate is collected.

When the analytes are retained, the technique has to retain the analytes on a stationary phase which is strong enough that the targeted sample components do not move through the packing material until the eluent is introduced. In such situations, the working procedure is more laborious, involving four stages (figure 65):

- conditioning;
- sample loading;
- interferences' elution (rinsing);
- analyte(s)' elution.

Conditioning is necessary because SPE cartridges are supplied dry and the stationary phase has to be wetted and equilibrated, using several bed volumes of wetting solvent (ultrapure water or low ionic strength buffer, as specified by the producer for each SPE type). The conditioning step establishes a dynamic equilibrium between the conditioning mixture and the packing material; in this stage, it is important that the wetting agent must not be allowed to dry since a SPE cartridge will dry in less than a minute of air passing through it when attached to a vacuum manifold.

Sample loading is accomplished to the top of the cartridge, using a low flow rate, to enable a proper interaction of the analytes with the packing material. Despite most of the unwanted components from the sample are washed away from the cartridge in this stage, important amounts remain trapped in the packing hence they have to be removed by **rinsing** (usually with ultrapure water or low ionic strength buffer). **The elution** of analytes requires an appropriate eluent (buffer with a high ionic strength), able to elute the compounds of interest in the smallest possible volume (1–2 mL). Detailed instructions are available in the technical descriptions from SPE's producers and these have to be followed rigurously to obtain relevant analytical results. Irrespective the stage, the flow rate of the eluent through a SPE cartridge should be in the range of ~1 mL/min or even lower; if the flow is too rapid, the sample components have insufficient time to interact properly with the stationary phase and losses occur.

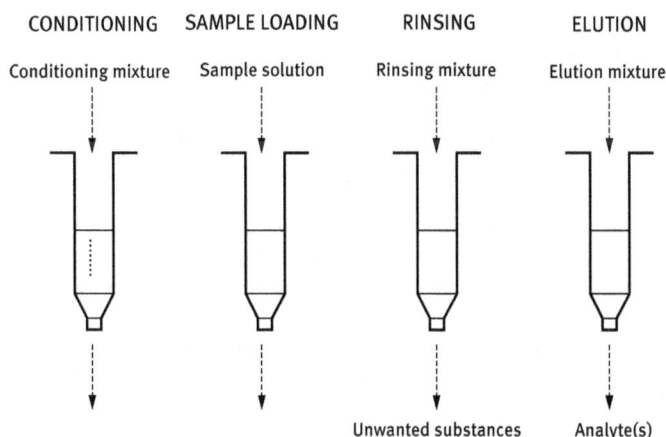

CONDITIONING   SAMPLE LOADING   RINSING   ELUTION

Conditioning mixture   Sample solution   Rinsing mixture   Elution mixture

Unwanted substances   Analyte(s)

Figure 65: Schematic of a generic SPE protocol which retains analytes.

Depending on the target objective of SPE, a proper packing type is necessary:

– reversed-phase packings are hydrophobic materials which are able to retain nonpolar compounds from a polar matrix while the polar analyses pass through unretained;
– ion-exchange resins retain charged compounds.

SPE is useful especially in cases in which the interfering compound and the analytes are both soluble and:

– the analytes are present at very low concentrations;
– there is a big difference in the analytes' concentrations;
– the samples contain hydrophobic components which may be irreversibly retained on the column, shortening the column lifetime.

In fact, SPE allows a wide range of sample-related operations, such as purification, preconcentration, fractionation and desalting. For more than three decades, SPE has become one of the fastest growing sample preparation and clean-up techniques for IC analysis, being also available in automated systems (Fan and Zhou, 2007; Frenzel and Michalski, 2016; Henderson et al., 1991; Montgomery et al., 1998; Pereira, 1992; Saari-Nordhaus et al., 1992; Seubert et al., 2004; Sun et al., 2006; Wang et al., 2014). If performed correctly, SPE can yield very clean extracts (hence membrane filtration is not necessary after this stage), being capable of concentrating significantly the target analytes.

References

Buldini, P. L., Mevoli, A., & Quirini, A. (2000). On-line microdialysis–ion chromatographic determination of inorganic anions in olive-oil mill wastewater. Journal of Chromatography. A, 882(1–2), 321–328.

Buszewski, B., & Szultka, M. (2012). Past, present and future of solid phase extraction: A review. Critical Reviews in Analytical Chemistry, 42(3), 198–213.

Campbell, I. (2012). Biophysical techniques. Oxford University Press, New York.

De Borba, B. M., Brewer, J. M., & Camarda, J. (2001). On-line dialysis as a sample preparation technique for ion chromatography. Journal of Chromatography. A, 919(1), 59–65.

Fan, Y., & Zhu, Y. (2007). Sample pretreatment in ion chromatography. Chinese Journal of Chromatography, 25(5), 633–640.

Frenzel, W., & Michalski, R. (2016). Sample preparation techniques for ion chromatography. In Michalski, R. (Ed.) Application of IC-MS and IC-ICP-MS in Environmental Research, John Wiley & Sons, Hoboken, New Jersey. 210–266.

Frenzel, W., & Markeviciute, I. (2017). Membrane-based sample preparation for ion chromatography – Techniques, instrumental configurations and applications. Journal of Chromatography. A, 1479, 1–19.

Fritz, J. S. (1999). Analytical solid-phase extraction. New York, Wiley-VCH.

Gandhi, J. C. (2003). Dialysis -a unique proven sample preparation technique for ion chromatography. LC-GC North America, 21(2), S40–S40.

Haddad, P. R., Doble, P., & Macka, M. (1999). Developments in sample preparation and separation techniques for the determination of inorganic ions by ion chromatography and capillary electrophoresis. Journal of Chromatography. A, 856(1–2), 145–177.

Henderson, I. K., Saari-Norhaus, R., & Anderson, J. M. Jr (1991). Sample preparation for ion chromatography by solid-phase extraction. Journal of Chromatography. A, 546, 61–71.

Li, H., Jing-Ping, L., & Xiao-Wen, R. (2008). Direct determination of choline in milk powder for infant and young children by online dialysis-ion chromatography. Chinese Journal of Health Laboratory Technology, 18(3), 444–445.

Liu, J. M., Liu, C. C., Fang, G. Z., & Wang, S. (2015). Advanced analytical methods and sample preparation for ion chromatography techniques. RSC Advances, 5(72), 58713–58726.

Lundanes, E., Reubsaet, L., & Greibrokk, T. (2013). Chromatography: Basic principles, sample preparations and related methods. John Wiley & Sons, Hoboken, New Jersey.

MISP – Metrohm inline sample preparation. Metrohm, Herisau, Switzerland (Accessed April, 26, 2022, at https://www.metrohm.com/content/dam/metrohm/shared/documents/brochures/89xx/89405002EN.pdf)

Mohammad, A. W., Ng, C. Y., Lim, Y. P., & Ng, G. H. (2012). Ultrafiltration in food processing industry: Review on application, membrane fouling and fouling control. Food and Bioprocess Technology, 5(4), 1143–1156.

Moldoveanu, S. C., & David, V. (2021). Modern sample preparation for chromatography. Elsevier, Amsterdam.

Montgomery, R. M., Saari-Nordhaus, R., Nair, L. M., & Anderson, J. M. Jr (1998). On-line sample preparation techniques for ion chromatography. Journal of Chromatography. A, 804(1–2), 55–62.

Nordmeyer, F. R., & Hansen, L. D. (1982). Automatic dialyzing-injection system for liquid chromatography of ions and small molecules. Analytical Chemistry, 54(14), 2605–2607.

Pereira, C. F. (1992). Application of ion chromatography to the determination of inorganic anions in foodstuffs. Journal of Chromatography. A, 624(1–2), 451–470.

Pomeranz, Y. Ed., (2013). Food analysis: Theory and practice. Springer Science & Business Media, Berlin.

Poole, C. F. (2002). Principles and practice of solid-phase extraction. Comprehensive Analytical Chemistry, Elsevier, 37, 341–387.

Rick, S., Steinbach, A., & Wille, A. (2009). Analysis of food samples with ion chromatography after inline dialysis. In Conference Proceedings, LC-GC, Europe (Accessed May, 17, 2022, at https://www.researchgate.net/publication/260302301_Analysis_of_Food_Samples_with_Ion_Chromatography_After_In-line_Dialysis).

Saari-Nordhaus, R., Nair, L. M., & Anderson, J. M. Jr (1994). Elimination of matrix interferences in ion chromatography by the use of solid-phase extraction disks. Journal of Chromatography. A, 671(1–2), 159–163.

Seubert, A., Frenzel, W., Schäfer, H., Bogenschütz, G., & Schäfer, J. (2004). Sample preparation techniques for ion chromatography. (Accessed July, 10, 2022, at https://www.metrohm.com/content/dam/metrohm/shared/documents/monographs/80255001DE.pdf).

Simpson, N. J. (2000). Solid-phase extraction: Principles, techniques and applications. CRC Press, Boca Raton.

Slingsby, R., & Kiser, R. (2001). Sample treatment techniques and methodologies for ion chromatography. TrAC Trends in Analytical Chemistry, 20(6–7), 288–295.

Song, W., Zhang, L. H., Lu, Y. F., & MA, Y. J. (2006). On-line dialysis-ion chromatographic determination of bromate in flour and its products. Physical Testing and Chemical Analysis/ Part B – Chemical Analysis, 42(11), 899–902.

Sun, Y., Huang, J., & Gu, P. (2006). Determination of trace haloacetic acids in drinking water using ion chromatography coupled with solid phase extraction. Chinese Journal of Chromatography, 24(3), 298–301.

Thomas, S. L., Thacker, J. B., Schug, K. A., & Maráková, K. (2021). Sample preparation and fractionation techniques for intact proteins for mass spectrometric analysis. Journal of Separation Science, 44(1), 211–246.

Vérette, E., Qian, F., & Mangani, F. (1995). On-line dialysis with high-performance liquid chromatography for the automated preparation and analysis of sugars and organic acids in foods and beverages. Journal of Chromatography. A, 705(2), 195–203.

Wang, N., Shou, D., Zhang, J., Li, G., Zhang, P., & Zhu, Y. (2014). Recent applications of sample pretreatment techniques in food analysis by ion chromatography. Journal of Food Safety and Quality, 5(5), 1287–1296.

Wiśniewski, J. R. (2019). Filter aided sample preparation–a tutorial. Analytica Chimica Acta, 1090, 23–30.

Yao, J., Hang, Y. P., Zhong, Z. X., Chen, M., & Li, M. (2010). Simultaneous determination of nitrite and nitrate in cooked meat by on-line dialysis-ion chromatography. Food Science, 31(2), 187–190.

Żwir-Ferenc, A., & Biziuk, M. (2006). Solid phase extraction technique-trends, opportunities and applications. Polish Journal of Environmental Studies, 15(5), 677–690.

6 Troubleshooting issues

In order to have a reliable and trouble-free IC system, certain issues have to be addressed properly and a good maintenance is a must (Foster, 2005; Hong and Fountain, 2012; Weiss, 2016). Some points to follow are:

- the mobile phase has to be prepared accurately, using high purity components, then filtered and degassed before use;
- the mobile phase reservoir has to be monitored to prevent running out the content;
- the pumping system must not be allowed to run dry (such situation can cause an exaggerate wearing of the seals) and has to be regularly checked to observe leakages when they occur, when the change of seals are necessary (or better to replace seals on a regular, programmed basis);
- the purging procedure has to be performed only after opening the purging valve;
- the samples have to be carefully filtered before injection;
- when using manual injectors, injection has to be accomplished by introducing the sample in the injection valve having the valve in the "open" position; at least three loop volumes are necessary for filled loop injections and air has to be carefully excluded from the syringe;
- when an autoinjector is used, it is necessary to use compatible vials for it, to fill the vials with proper volumes and to attach firmly the caps;
- the column is the place in which separation occurs, hence it has to be handled with great care: the fittings have to be compatible with it and not overtighten, only compatible mobile phases have to be used;
- when not in use, a column has to be stored in the recommended mobile phase, with plugs well fitted in both ends and never allowed to dry out;
- the chromatographic data system provides in most cases warnings for different problems detected in the system; these warnings should be solved in due time in order to have a proper system operation;
- the waste container has to be monitored on a daily basis to prevent overflowing, especially during long unattended runs.

A proper operation can only be accomplished after the system's equilibration with the desired mobile phase and flow rate, at the temperature specified in the method, until a stable baseline is achieved.

It is recommended to keep a journal for each column, recording the mobile phases used, flow rate, backpressure, the number and type of samples injected, observations regarding changes in peak shapes, retention time shifting, any sign of contamination, etc. For traceability, it is also important to record the actions performed during maintenance and troubleshooting.

https://doi.org/10.1515/9783110644401-006

6.1 Baseline problems

An ideal baseline is straight and free from noise; however, it can be affected by the mobile phase flow and/or by environmental factors; when irregular shapes can be observed, the most common problems are drifting (usually a gradual increase of the baseline level over time) and noise (Katsumine et al., 1999).

Drifting is a common issue when the system is not left enough for a proper warming-up and equilibration. It can have many possible causes:

- changes in the eluent temperature;
- changes in the eluent composition;
- changes in the flow rate;
- certain contaminants washed out from the column;
- detector lamp warming up.

Noise is a random change of low amplitude and high frequency in the chromatogram, being the result of the summation of several components:

- electric noise (comes from the electricity supply and from electronic components);
- pump noise (corresponds to the opening and closure of the check valves), it changes with the flow rate and stops when the pumping system stops;
- the presence of air in the detector's flow cell.

Such issues can be addressed by acting against the supposed cause, hence it is important to ensure enough time for system warm up, to check for leaks and to check the proper function of the degasser module. If the valves are old and worn out, they have to be replaced. If fluctuations appear to be synchronous with pumping system's cycle, the pumping system has to be purged.

6.2 Retention time reproducibility

Retention time is a parameter of utmost importance in IC, being related with both qualitative analysis (because the chromatographic peaks are identified by the chromatographic data system by their retention times) and quantitative analysis (as peaks' retention times change so does their area). If retention times change, the most common outcome is incorrect peak assignment or peaks being missed. Among the main causes of unwanted changes in retention times are changes that occur in temperature, composition of the mobile phase, flow rate, sample overloading, as well as column wearing (Dolan, 2016; Fritz, 2005; Zang et al., 2021).

Temperature changes leads to changes in both retention times and in selectivity. Because laboratory room temperatures can vary by several degrees during the course of one day as a result of air conditioning, heating, changes of environmental temperature or even air currents caused by open doors or windows, maintaining a constant temperature during IC separations is very important and having a perform-ant column oven in the IC system configuration is a must.

Flow rate is directly related with retention time; at a given and constant reten-tion time, peaks elute at a given retention time, and if the flow rate changes, the retention time also changes. Flow rate changes can occur due to the presence of air bubbles in the pump head, faulty check valve(s), leaks or operating at the extremes of a pump flow rates' range.

Changes in the mobile phase composition can be caused by faulty check valve(s), plugging of the inlet filter from the mobile phase reservoir, poor mixing or evapora-tion of volatile components. Preparing of a fresh mobile phase or the change of the inlet filter can solve the problem.

Overloading occurs when injecting more sample than a chromatographic col-umn can process; in such a situation, the resulting peaks will elute later than ex-pected, will be broader, eventually with more tailing. Such a problem can be solved easily by injecting a smaller volume of sample or by diluting the sample.

Column wearing (especially due to contamination) usually causes not only changes in retention times but also a decrease in resolution and a higher backpres-sure. Column usage can be the cause for a decrease in retention because irreversible adsorption of certain species will reduce the column's capacity. Column rinsing or column regeneration is a possible solution in such cases, when the manufacturer's recommendations have to be followed (see also Section 6.8).

6.3 Peak height and peak area reproducibility

If by repeated injections of the same sample significant changes occur in peaks' height and peak areas, while the retention times are not affected, quantitative re-sults are distorted since the sample concentration should be directly proportional to peak area and peak height. Such problems can arise either because a different amount of sample has been injected, or flow rate varies or because the system is responding differently (Bidlingmeyer, 2002; Dolan, 2011a; Stoll, 2019).

When using manual injection, air bubbles can be sucked in the syringe and/or in the injector, the needle position can be wrong (not engaged in the sleeve from the rotor seal) or an insufficient volume can be injected when using complete loop filing (less than three times the loop volume). When performing a partial loop filling, the injected volumes usually vary because they depend on the operators' ability to inject the same volumes each time. When using autoinjectors, variations in peak area/peak height can be due to the presence of some air in the syringe, partially blocked needle, low volume of sample, inaccessible to the needle position setting or due to an im-proper washing procedure between injections. In the case of a partially blocked nee-dle (e.g., with particles from the sample), samples are not sucked up properly from the vials, so very small peaks or quite no peaks result; by turning off the autosampler, then removing the needle, sometimes, it is possible to rod it out with a fine needle cleaning wire (if this is not working, it is necessary to replace the needle). Any air bubble trapped in the syringe or in the injector results in a random variation of the injection volume from an injection to another, hence to different peak areas.

Variations in the flow rate can also be a cause of peak area/peak height variability; they occur as a result of check valves' malfunctioning, of air trapped in the pump head or of leaks; changes in retention times occur in such situations as well. Monitoring the backpressure is a useful diagnostic tool to identify the problem; leaks are usually easy to find.

The detector can be a cause for changes in sensitivity (e.g., if a detector lamp – such in UV-VIS or PDA detectors – is getting old, the peaks get smaller).

6.4 Ghost peaks

Ghost peaks appear as normal peaks in a chromatogram but occur without having a correspondent analyte in a sample. There are three major causes for the appearance of ghost peaks (Dolan, 2013; Williams, 2004; Zhang et al., 2011):
–   contamination (e.g., from an improperly washed sample loop from the injector);
–   late eluting analyte(s) from a previous injection, which can arise because the run time for a separation is too short and one or more strongly retained compound(s) appears unexpectedly on a chromatogram or because the system has not been allowed to reach equilibrium;
–   impure substances used in the mobile phase's preparation (if a component of the mobile phase has a certain retention on the column, it will show up as a retained peak).

The easiest way for getting rid of such peaks is to increase the run time and/or the flow rate; also, it is compulsory to clean-up properly the syringe used for injections. A typical source of ghost peaks is the ultrapure water's quality; this water can contain impurities from different sources (e.g., the water purification system itself, the container in which water is stored, bacterial growth), hence the performance of the purification system has to be checked on a regular basis (Dolan and Snyder, 1989).

6.5 Changes in peak shapes

A common chromatographic peak has an almost Gaussian shape, with equal slopes to both leading and trailing edges; when changes in shape occur, asymmetrical or broad peaks result, which can lead to inaccurate quantifications. Common causes of changes in peak shapes are: column overload, column wearing (mostly because of chemical effects), dead volumes at the column head, coelution of compounds with similar chromatographic behavior, improper injection technique or even leaks (Bidlingmeyer, 2002; Dolan and Snyder, 1989; Hong and Fountain, 2012; Meyer, 2013).

If the injected amount of analyte(s) is too large, exceeding the maximum loading of a column, broad asymmetric peaks with severe tailing are obtained. This condition is called overloading and is due to the fact that each column packing has a certain number of ion-exchange sites, which are occupied by the eluent's ions before a sample is injected; after injection, the ion-exchange process between the eluent's ions and sample's ions lead to the desired separation. The exchange process works properly only when the number of exchange sites is higher than the number of bonding places required by the analytes from the sample; otherwise, a column overload occurs, leading to cut-off peaks, shark-fin-shaped peaks, fronting peaks, tailing peaks, altered retention times, but all of these can be solved easily by diluting the sample and reinjecting it or injecting a lower sample volume.

For changes in peak shapes due to the other above-mentioned causes, the remedies are more complex (e.g., a column void can be eventually refilled and reassembled with a new frit, as indicated in the later subchapter of this section, while resolving an interfering peak needs important changes in the mobile phase composition). An increase in the oven temperature can help in the case of broad peaks which can become narrower. Extra-column effects can also contribute; they can be reduced by replumbing the system with shorter and narrower tubing. Changes in peak shapes or tailing make quantification more difficult and is often used as a criteria for the end of useful life of a separation column.

6.6 Pressure issues and leaks

Monitoring backpressure in IC systems is a fast and useful diagnostic tool in identifying numerous problems encountered in practice; to be effective, it is important to have a journal in order to start troubleshooting from a reference situation. Table 16 presents common pressure problems, compiled from several references (Dolan, 2011b; Dong, 2006; McMaster, 2007; Meyer, 2013b; Raval and Patel, 2020).

Besides increased mobile phase consumption, leaks cause serious technical problems, leading finally to possible incorrect peak assignments and incorrect quantitation (both issues being related with the flow rate of the mobile phase). A proper knowledge of possible causes of leaks allows us to adopt measures to prevent them. Leaks are common in high-pressure systems, the possible causes being quite diverse, some of them being highlighted in Table 17 (Dolan, 2011; Dong, 2006; McMaster, 2007; Meyer, 2013b; Raval and Patel, 2020; Runser, 1981).

Leaks from the pumping system have the greatest likelihood manifesting themselves because there the system is under the highest backpressure; usually a puddle under the pumping system can be observed. The most obvious source of leaks in a

Table 16: Common pressure problems.

Problem	Cause	Remedy, prevention
No pressure/ no flow	No mobile phase	Reprime the mobile phase reservoir
	Air in the pump head	Degas the mobile phase, purge the pump
	Faulty check valve(s)	Replace check valves
	Broken piston	Replace piston and piston seal
High backpressure	Plugged column inlet frit	Backflush column or replace frit or replace column
	Plugged in-line filter or guard column	Replace in-line frit or guard column
	Plugged injector	Clean or replace injector
	Plugged tubing	Replace tubing after systematically disconnecting system components from the detector end to the blockage (for localizing the plugged tube)
	High viscosity of the mobile phase	Increase the temperature
	Microbial growth in the column	Use a mobile phase with 10% methanol or acetonitrile; prepare mobile phase daily
Low pressure	Mobile phase leak	Check flow path connections from pump to column outlet; tighten all fittings, replace defective fittings
Pressure fluctuations	Air in the pump head	Degas the mobile phase, purge the pump with degassed mobile phase
	Faulty check valve(s) – not sitting properly/dirt or air in check valve	Disassemble and clean (by sonication in water: methanol) or replace check valves
	Pump piston seal failure	Replace the defective seal
	Salt deposits	Wash with water, replace seal if necessary
	Mobile phase leak	Locate the leaking source and resolve it

pump is the piston seal and there are two approaches in changing: preventive change (at least once per year) or change as they start to leak (a cheaper option since a seal can last several years).

Table 17: Common leaking problems.

Leak source	Cause	Remedy
Tubing	Improper bending of PEEK tubing, high pressure in the system	Replacing the leaking piece
Fittings	Loose fittings	Tighten carefully
	Mismatches parts	Use parts from the same brand
	Dirty fitting	Disassemble and clean
Pumping system	Loose check valves or fittings	Fully tighten
	Pump piston seal failure	Replace piston seal
Injector	Rotor seal failure	Replace rotor seal
	Loose injection port seal	Tighten carefully
Column	Loose end fitting	Tighten carefully
	Blocked end frit	Replace end frit
Detector	Cracked flow cell window	Replace window/flow cell
	Blocked waste line	Replace waste line

6.7 Autosampler issues

As mentioned in Section 2.3.4, the presence of an autosampler in an IC is beneficial, but in the meanwhile add more complexity and requires more attention from the user. The autosampler requires one more reservoir, beside that/those used for the mobile phase(s): a "wash bottle," usually filled with ultrapure water, which is necessary for rinsing the injection port and the injector's needle; this also requires a proper attention since if it becomes empty, then the injection process will be compromised. Some common problems occurring when working with autosamplers are summarized in Table 18 (Dolan, 1987; Dong, 2006; Paul et al., 2019; Raval and Patel, 2020).

A careful labeling of sample vials, then an attentive correspondence between the position of each sample in the sample rack and the sample's ID in the analytical batch procedure is a must; otherwise, if a vial is in an incorrect location, either unexpected peaks or no peaks will result/if the vial range is incorrect, then the wrong samples will be injected and the obtained results will be meaningless. Each batch has to be carefully supervised to reflect the sample ranges to be analyzed.

Table 18: Common sample injection problems when using autosamplers.

Problem	Cause	Remedy
No peak	Trapped air	Purging the autosampler
	No sample vial in the position	Rearrange the sample vials in the autosampler's rack
	Improper filling of a sample vial (the sampling needle cannot reach the sample)	Fill properly the sample vial or change the autosampler's settings
Poor precision	Air bubbles	Purging the autosampler
	Worn out syringe, rotor or seal	Replace the defective component
High backpressure during injection	Salt deposits on the valve and /or on the loop	Dismantle and sonicate, replace the rotor/seal or loop if necessary
Leaks	Worn or scratched injector valve rotor, damaged seals	Replace the defective component(s)

6.8 Column regeneration

Improper use or common wear process of a chromatographic column will lead to a decrease of performance, broad or split peaks, asymmetric peaks, changes in retention times, high back-pressure; the quality of the separations will generally be poorer. The most common causes of column failure are: plugging the column with particulates, adsorption of impurities and mechanical shocks (Majors, 2003).

Particulates often originate from the sample or from the mobile phase, but can also be generated by the common wear process of certain IC system components, such as pumps or the moving parts of the injector, which yield particles that can be trapped to the column inlet. To avoid column plugging with particulate matter, samples should be filtered using 0.47 µm membrane filters; mobile phases containing dissolved solids should also be filtered before use, while clean glassware has to be used in all operations and the mobile phase reservoirs are properly covered to keep out the dust. For protecting the columns from particulates originating from system's wear, in-line filters placed between the sample injector and the column inlet are the best solutions.

Retention of certain compounds from complex samples can gradually accumulate on the stationary phase from the column, leading to change in chromatography (changes in retention times, increase in peak widths, loss of efficiency). To prevent such problems, an initial cleanup of the sample using proper sample preparation techniques before injection is a must.

Mechanical shocks will not brake an IC column but can act on the column's packing, disturbing the particles, with the formation of a space (or "void"), usually at the column's inlet; a void will cause in most cases an important loss in column performance. Voids can also be caused by big pressure changes (e.g., pressure variations caused by an improper injection).

Before abandoning a problematic column, it is worthwhile trying to clean or to regenerate it.

An increase in system backpressure can be a symptom of such a situation; a higher back-pressure can have as causes a plugged frit, a plugged column bed or even a contaminated column bed. Provided that the peaks' shapes are not distorted, a simple test for a plugged frit is to reverse the column position in the system without connecting the detector, then starting the pumping system and observing the back-pressure: if it drops significantly, then the inlet frit is plugged, but if drops only a little then remains high, the column bed is blocked.

It is possible to replace a plugged frit with a new one, supplied by the column's manufacturer, after clamping the column firmly in a vice to prevent uncontrolled movements, then carefully dismounting the inlet end-fitting, removing the plugged frit and replacing it with the new one and finally reassembling the end-fitting; this replacement has to be accomplished with utmost care, preventing any disturbance of the stationary phase, since it can lead to column's deterioration. If after backflushing the column's performance is still bad, a regeneration procedure may help.

If the peak shapes are distorted, it is possible to have a void in the top of the column and reversing the column is not advisable; void refilling is possible working in a similar manner as for frit replacement; after removing the frit, it is necessary to add a slurry of the same stationary phase as the one from the column with the aid of a small spatula, up to the void's filling, then reattach the frit and reassembling the end-fitting. If not performed carefully, this procedure can compromise the column, which has to be replaced.

Sometimes, regeneration procedures are recommended by column producers in their datasheets. A common cleanup can be accomplished with 10× eluent strength/ 0.2 M HCl in 80% acetonitrile (HCl to remove metals, acetonitrile to remove organics), overnight, using small flow rates – 0.1 mL/min for 2 mm i.d. columns up to 0.25 mL/min for 4 mm i.d. columns.

Such corrective measures will not return a column to its original condition, but sometimes may help in performing more separations up to the moment a new column is provided. If the regeneration procedure fails, one has to remember that a column is a consumable; hence when its performance is degrading, it has to be replaced with a new one.

References

Bidlingmeyer, B. (2002). Liquid chromatography problem solving and troubleshooting. Journal of Chromatographic Science, 40(7), 417–418.

Dolan, J. W. (2016). Retention time drift – a case study. LC-GC Europe, 29(4), 206–212.

Dolan, J. W. (2013). Gradient elution, Part VI: Ghost peaks: Where do all those extra peaks come from?. LC-GC North America, 31(8), 604–609.

Dolan, J. W. (2011). Troubleshooting basics, Part I: Where to start? Where do you start when you encounter a new problem with your liquid chromatography system?. LC-GC North America, 29(7), 570–573.

Dolan, J. W. (2011b). Troubleshooting basics, Part II: Pressure problems. LC-GC North America, 29(9), 818–824.

Dolan, J. W., & Snyder, L. R. (1989). Troubleshooting LC systems: A comprehensive approach to troubleshooting LC equipment and separations. Springer Science & Business Media, New York.

Dolan, J. W. (1987). Troubleshooting: Troubleshooting autosamplers. II. LC-GC North America, 5(3), 224–226.

Dong, M. W. (2006). Modern HPLC for practicing scientists. John Wiley & Sons Inc. Hoboken, New Jersey.

Foster, K. L. (2005). Handbook of ion chromatography, 3rd edition. Wiley-VCH Verlag, Weinheim.

Fritz, J. S. (2005). Factors affecting selectivity in ion chromatography. Journal of Chromatography. A, 1085(1), 8–17.

Hong, P., & Fountain, K. J. (2012). Guidelines for routine use and maintenance of ultra performance size exclusion and ion exchange chromatography systems. (Accessed April 29, 2022, at https://www.waters.com/content/dam/waters/en/app-notes/2012/720004182/720004182-de.pdf)

Katsumine, M., Iwaki, K., Matsuda, R., & Hayashi, Y. (1999). Routine check of baseline noise in ion chromatography. Journal of Chromatography. A, 833(1), 97–104.

Majors, R. E. (2003). The cleaning and regeneration of reversed-phase HPLC columns. LC GC North America, 21(1), 19–27.

McMaster, M. C. (2007). HPLC: A practical user's guide. John Wiley & Sons, Hoboken, New Jersey.

Meyer, V. R. (2013). Pitfalls and errors of HPLC in pictures. Wiley-VCH, Weinheim.

Meyer, V. R. (2013b). Practical high-performance liquid chromatography. John Wiley & Sons, Chichester.

Paul, C., Steiner, F., & Dong, M. W. (2019). HPLC autosamplers: Perspectives, principles and practices. LC-GC North America, 37(8), 514–529.

Raval, K., & Patel, H. (2020). Review on common observed HPLC troubleshooting problems. International Journal of Pharma Research and Health Sciences, 8(4), 3195–3202.

Runser, D. J. (1981). Maintaining and troubleshooting HPLC systems: Users guide. John Wiley & Sons, New York.

Stoll, D. (2019). What's trending in LC troubleshooting? LC-GC Asia Pacific, 22(1), 14–17.

Weiss, J. (2016). Handbook of Ion Chromatography, John Wiley & Sons Hoboken, New Jersey.

Williams, S. (2004). Ghost peaks in reversed-phase gradient HPLC: A review and update. Journal of Chromatography. A, 1052(1–2), 1–11.

Zang, W., Sharma, R., Li, M. W. H., & Fan, X. (2021). Retention time trajectory matching for target compound peak identification in chromatographic analysis. ArXiv Preprint arXiv:2107.07658. (Accessed April 29, 2022, at https://arxiv.org/ftp/arxiv/papers/ 2107/2107.07658.pdf)

Zhang, D. D., Sadikin, S., Redkar, S., & Inloes, R. (2011). Ghost peak investigation in a reversed-phase gradient LC system. LC GC North America, 29(5), 394–400.

Index

absorbance 1, 6, 19, 57, 65, 68–70
absorbance units 65
adjusted retention time 9
adsorption chromatography 3–4
adulteration 128–129, 142
alcoholic beverages 130
amperometric detection 73
antioxidant 110
arsenate 62, 89, 93, 144
asymmetry factor 7–8, 28
authenticity 107–108, 128–129, 137, 142
autoinjectors 46
auxiliary electrode 73

bacterial growth 38, 43, 166
baseline 11–12, 23–24, 27, 37, 39, 46, 55,
    57–58, 72, 139, 153, 163–164
baseline width 11–12, 27
Beer 98
beer wort 99
bench footprint 79
Biogenic amines 74, 107–108, 114–115
block heathers 54
bromate 62, 67, 73, 88–89, 92, 139, 158
brominated vegetable oils 147
burtonization 100

calibration 19, 21, 23
calibration curve 21–22, 26, 57
capacity factor 9–10, 13
capillary columns 52, 140
capillary IC V, VI, 137, 140
carbohydrate analyzer 79
CDS 76–78
charge detection 141
chemical suppression 55–56
chlorination 92
choline 64, 120, 124
chromate 62, 66, 89, 93
chromatographic data system 8, 16, 19, 21, 34,
    42, 46, 76, 155, 163–164
chromatographic system 3, 5–6, 8, 10–12,
    23–25, 27, 36, 44, 46, 52, 155
chromatography 2–3, 31, 78–79, 147
chromophore 66
co-chromatography 17–18
coelutions 14

coffee 128, 142, 144
column ovens 54
column wearing 165
combustion IC V, VI, 137, 146–147
combustion ion chromatography 146
compression screws 49
conditioning 159
contamination 25–26, 53, 92, 108–109, 137,
    139, 156, 163, 165–166
corn stover 144
counterion 24

dead time 8–9, 10, 14, 24, 29
degassing 24, 39, 91, 102–105
dialysis 157
dilution VI, 2, 28, 35, 46, 91, 102–106, 109,
    122, 124–125, 128, 153–155
diode array detectors 69
direct IC 37, 55
direct UV-VIS detection 66
distribution coefficient 5
drift 24, 27, 58, 72, 78
drifting 164
dynamic range 1–2

efficiency IX, 12–13, 14, 27–28, 51–52, 54, 58,
    78, 119, 137, 140, 153, 170
EG 138, 140
electrical conductivity 58
electrochemical detectors 73
electrolysis 137, 139
electronic suppression 55
eluent generation V, VI, 137
elution 3, 7, 25, 37, 41–42, 57, 60, 70–72, 74,
    88, 100, 146, 159
environmental-friendly 1
equivalent conductivity 59
ethanol 75, 98–99, 102, 107, 113,
    122, 133–134
external standard calibration method 19

fast screening 63, 108, 113
ferrules 49
filed loop injection 45
fingerprinting 129, 142
fittings 28, 40, 47–48, 50, 163,
    168–169

https://doi.org/10.1515/9783110644401-007

www.ingramcontent.com/pod-product-compliance
Lightning Source LLC
Chambersburg PA
CBHW081529220326
41598CB00036B/6378